D0571860

# Energy Aftermath

# Energy Aftermath

Thomas H. Lee
Ben C. Ball, Jr.
Richard D. Tabors

Harvard Business School Press
Boston, Massachusetts

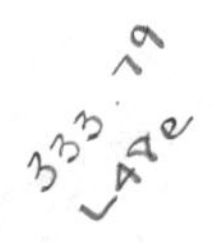

**Library of Congress Cataloging-in-Publication Data**

Lee, Thomas H., 1923–
Energy aftermath / Thomas H. Lee, Ben C. Ball, Jr., Richard D. Tabors.
p. cm.
Includes bibliographical references.
ISBN 0–87584–219–4 (alk. paper)
1. Energy policy—United States. 2. Energy industries—United States. 3. Energy consumption—United States. 4. Energy policy. 5. Energy industries. 6. Energy consumption. I. Ball, Ben C., 1928– . II. Tabors, Richard D. III. Title.
HD9502.U52L44 1990
333.79′0973—dc20 89–39842
CIP

The paper used in this publication meets the requirements of the American National Standard for Permanence of Paper for Printed Library Materials Z39.49-1984.


Printed in the United States of America
94 93 92 91 90 5 4 3 2 1

To Our Wives
Kin Ping, Helen, and Patton

# Contents

# Preface

As the final editing is completed for *Energy Aftermath*, the urgency of its message seems to be increasing. Whatever the outcome of the current (summer 1989) furor over "cold fusion in a jelly glass," the wrong questions are still being asked. For example, "Is it or is it not *the solution*?" We are still mistaking science for technology, overlooking economics, ignoring the dynamics of the commercialization of new technologies, hoping for energy substitution rather than technological substitution, and wishing for a miracle.

The debate over the Seabrook (New Hampshire) nuclear plant continues without engaging the fundamental issues of real economics, safety, robustness, waste disposal, eventual facilities retirement, and the environment.

Investors are lined up at courthouses, suing securities salesmen in an attempt to recover losses incurred in petroleum exploration and production ventures that turned out to be noneconomic. Many exploration and production companies and partnerships have simply gone bankrupt. The major oil companies are writing off billions of dollars of 1970s' exploration and production ventures that are now recognized to have ranged from the noneconomic to worthless. In a sense, such ventures contributed to their own demise. They were a by-product of the large increase in production capacity, which led to the softening of prices, which led to the poor economic results.

The largest oil company in the world ran one tanker aground in Alaska and the price of oil shot up almost as if there were another "embargo."

So, on the one hand, energy prices are down, finally. But, on the other hand, we are still nervous, are still paying the price of the mistakes of the last decade and a half, and are still reacting impulsively to all kinds of news from the energy front.

In our book we point out that many of the energy lessons are generic. They apply equally well to other sectors. Coincidental with this writing is the release of the MIT Commission on Industrial Productivity (Dertouzos, Lester, and Solow, 1989). Although that study did not explicitly include the energy sector, it arrived at conclusions that are similar to ours. First, the problems with current U.S. industrial strategy and practice are systemic, and largely the result of using outmoded paradigms. The *Boston Globe*'s (1989) characterization that ". . . new failures . . ." are ". . . tied to old successes" is accurate. Second, these problems can be corrected by thoughtful attention to the lessons that can be learned, and conscientious rebuilding of the systems that the MIT report concentrated on—corporate America. We believe the systems must include the financial institutions as well.

Selecting the title for a book is in a sense the most difficult part of writing it. The title must be both attractive and descriptive. We trust ours is attractive. The kind of events mentioned above convince us that it is fundamentally descriptive.

The usual meaning of "aftermath" is the results or consequences of something (Funk & Wagnall, 1988). Our use, however, is much richer than that. The word is derived from the Old English *mæth*, which refers to the cutting of grass.[1] Therefore, "aftermath" really means "the second harvest." During the 1970s and the 1980s, we had the first harvest of the energy crisis, which included wild uncertainty, gyrating prices, combined with unresolved analyses and conflicting syntheses.

This book deals with the energy crisis' second harvest. We are sure the energy situation remains subject to turmoil and serious discontinuities. The earlier crop has been mowed, although we are still paying for it.

If we will take seriously the lessons to be learned from the past, the "result or consequences"—the second harvest, *The Aftermath*—need not be "esp. an unpleasant one." It could well be a hopeful one.

Cambridge, Massachusetts
Summer 1989

Thomas H. Lee
Ben C. Ball, Jr.
Richard D. Tabors

## NOTE

1. af-ter-math n. [after + dial, math <OE, *mæth,* cutting of grass < mawan, to mow, with *-th* suffix] 1. a second crop, as of grass that grows after the earlier mowing 2. a result or consequence, esp. an unpleasant one. (Guralnik, 1974.)

## REFERENCES

*Boston Globe,* 1989. Wednesday, May 3, p. 16.

Dertouzos, Michael L., Richard K. Lester, and Robert M. Solow. 1989. *Made in America: Regaining the Productive Edge.* Cambridge, MA: MIT Press.

*Funk & Wagnall's Standard Desk Dictionary.* 1988. New York: McGraw-Hill.

Guralnik, David B., Editor in Chief. 1974. *Webster's New World Dictionary of the American Language.* New York: William Collins & World Publishing, p. 24.

# Acknowledgments

The idea for this book first saw the light of day when the four of us discussed it in a hotel room in Taipei in January 1986. Ed Schmidt was the fourth. Except for his untimely death in 1987, he would have been a coauthor. We are grateful to him for helping give birth to the project. Many of the ideas that we have incorporated are his. We hope we have done them justice.

We were in Taipei to participate in the meeting of the International Consortium for Integrated Energy Systems. Participation in this consortium has served as a fertile atmosphere in which to develop our concepts, and we appreciate the stimulation that our consortium colleagues provided.

We especially are indebted to our editor at the Harvard Business School Press, Richard A. Luecke. Not only has Dick improved the style and organization of the work; he has also made significant contributions to its content. Working with him we have learned that the secret to turning a manuscript into a good book is having an editor who is not only professionally competent but also collegial, supportive, and constantly encouraging. He has shown us that rewriting can be significantly productive, and therefore *almost* fun.

Claudia Staindl of the International Institute for Applied Systems Analysis (IIASA) provided valuable assistance during the first two years of our work by typing some early drafts and collecting many of the references.

The writing took place largely in Cambridge and Vienna. We thank MIT and IIASA for providing an environment conducive to this kind of work. We also drew on nearly half a century of personal experience in the energy sector at GE, Gulf Oil, MIT, IIASA, Ball & Associates, The MAC Group, and Tabors Caramanis & Associates. Without such experiences we could never have written *Energy Aftermath*.

# Introduction

As hard as it may be to recall, as recently as seventeen years ago "energy" was a transparent sector of the economy and society. For all its importance to the way we lived, energy was invisible, and never got in our way. Energy was something we studied in high school physics. Only those who made their living from it read books about it. Like the air we breathe, we took energy for granted.

One reason is that no one buys or uses energy as an end in itself. *All* demand for energy is *indirect,* and is derived only from the benefits it provides. People do not buy gasoline because they want gasoline, but because their cars need it to get them from point A to point B. People only purchase oil to keep their homes warm in the winter. The same is true in the industrial sector. A steel mill buys coal only because it needs it to make iron or steel. The same applies to electricity and natural gas.

Second, the price of all forms of energy was—until seventeen years ago—so low to the customer, relative to all other items, that energy cost was not a significant factor in decision making. The design of a home, office, or factory building was not determined or even seriously affected by the cost of heating or cooling it. In fact, the bulk of today's stock of residential and commercial buildings was designed as though energy costs mattered very little. Auto mileage was a topic of occasional idle conversation, but rarely a factor in the selection of a new model or in deciding to take a particular journey. Industries with high energy usage (e.g., chemical manufacture and metals smelting) used energy thoughtfully, but still their energy costs were small compared to other operating and capital costs. Oil companies and electric utilities were large, and had prominent roles as commercial and industrial firms. But despite the ubiquitousness of their products, the firms were not a source of interest to anyone outside the industries involved—with the exception of an occasional regulator or legislator.

Not only were energy prices low and even declining slightly in real terms, they also were stable and predictable. Energy demand growth was highly dependable, and supply was correctly assumed to be completely reliable. For example, despite economic cycles, the vagaries of the weather, and two traumatic closings of the then all-important Suez Canal, reliable supply and price were simply never visible issues. Even labor unrest in the industrialized world and political unrest in critical third world exporting countries did not disrupt either supply or price.

As everyone knows, that idyllic world came to an end—and not with a whimper. But that is not the problem. The problem is, when it ended, the energy experts led most people seriously astray and they continue to do so.

And not just small experts. Exxon, the world's largest corporation at the time, blundered into synfuels, along with the U.S. government's Synfuels Corporation and most of the major international oil companies; billions of dollars were wasted in that futile endeavor. Another of America's giants, Gulf Oil Corporation, opened and developed a major uranium mine in New Mexico, firm in the conviction that yellow cake (uranium oxide) would soon increase in price several times. Du Pont bought Conoco and U.S. Steel bought Marathon, both at prices significantly above market. The prestigious National Academy of Engineering conducted a study for the president under a rubric that proved to be a logical absurdity: "Project Independence."

In the years since the 1973 "oil shock," the U.S. government has wasted some $100 billion, and U.S. industry about an equal amount.

Long gone are natural gas shortages, when schools closed for lack of heat. With gasoline prices today as low in inflation-adjusted terms as they were in 1972, the price spikes of 1973–1974 and 1978–1979 seem like a bad dream. The picture of cars waiting in line for five gallons seems almost quaint. The odd day/even day rationing system, and the practice of service stations flying red flags to show they were "out of gasoline," even seem a little picturesque. One hears less and less about oil "shortages" and "gluts." Yankees worry little anymore about freezing in the dark, and the Texas-Louisiana Gulf Coast is recovering from its oil-bust. "Robber baron" oil companies and the "robber sheikhs" of OPEC are off the front page.

After only a decade of existence, the Department of Energy is beginning to look like a creature that has outlived its usefulness. There are no more debates about the "hard path" versus "the path not taken." It's been a couple of years since anyone has heard discussions of either *miraculum*, which would recover heat from the ocean, or *nonexistium*, which would allow cars to get a hundred miles per gallon. The younger generation will hardly know what their elders are talking about when they refer to the days when it looked like the end of the industrialization of the world was at hand, for the simple lack of energy.

The premise of our book is that it would be foolish to forget these fifteen or twenty years of energy chaos, without a diligent effort to understand what led up to it, identify the blunders made in responding to it, synthesize the lessons, and see what the future might hold.

Despite the dark days, the draconian fears, and the real damage done, the energy chaos was not the Great Calamity of the twentieth century. The United States has muddled through far greater traumas—wars, depressions, and natural disasters—and made perhaps even bigger mistakes and learned less. One school says, "Peoples and governments never have learned anything from history, or acted on principles deduced from it" (Hegel, 1832). This statement, however, should not comfort those responsible for analysis and for making decisions, past, present, or future—in the private or public sector. Perhaps, for the very reason that the energy chaos did not turn out to be the catastrophe it was feared to be, we might actually learn something from it, and act on principles deduced from it.

Our task is to develop the understanding, identification, and synthesis, and to suggest a direction for the coming decades. We adopted the following thesis in carrying out our task.

The events of 1973–1974, which included oil (and gasoline) shortages, oil price spikes, and an oil embargo, were and are termed an energy crisis—a misleading choice of terms. In fact, these events signaled not an occurrence but the beginning of a New Energy Era, one that the entire world would have to deal with for the foreseeable future. The new energy situation differs from that of the past in ways so fundamental that most of the given frameworks for thinking about energy became obsolete. Acknowledged experts failed to recognize the new era, and to appreciate how radically it differed from the past they understood so well. The result was massive confusion, misdirection, and wasteful efforts. More important, the results exacerbated the effects of the problem itself: Shortages were greater, price swings were more erratic, and uncertainty was increased.

As usually happens, the worst-feared scenarios did not occur, and the worst advice was not followed. We were spared some self-inflicted injuries. But the crisis is not over, despite today's apparent calm. We are still in the New Energy Era: an era of uncertainty and volatility, which decision makers must learn to face, and with which they must struggle over the coming decades.

Underlying much of our analysis are assumptions regarding the workings of the market, and the relation between the private and public sectors, particularly in the United States. Most energy decisions are made in the private sector, that is, through the workings of the market. We are under no illusions about that imaginary thing called the "perfect market." When we talk of the market, we mean the one that actually exists, with all of its imperfections.

When we say most energy decisions are made in the private sector, we are referring to the fact that the U.S. government—or, for that matter, the various states—are neither producers nor consumers of energy, neither sellers nor buyers of energy in truly significant quantities. Of course, there are exceptions, such as the TVA as a supplier of electricity, and the military as a consumer of fuels. As a percentage of total energy usage, however, the extent to which the government is involved as buyer or seller is quite small. The major and, we believe, the appropriate role of government in the existing market economy is to distance itself from active participation and instead provide the *context* for market decisions by the private sector. The government's role is not to make market decisions, that is, to set prices or allocate resources or goods in the private sector. Rather, its role is to allow market forces in the private sector to make the decisions about prices and the distribution of resources and goods. Nevertheless, it remains the government's responsibility to *influence* private-sector decisions so that public policy is served. For example, a tariff is justifiable in this view if domestic production of the item is of greater value to the society than an imported item by approximately the amount of the tariff. Then, given the context of that tariff, the private-sector market would determine how much is produced, how much is consumed, how much is imported, and at what prices.

In providing the context for energy-market decisions, the government has traditionally been quite active. Examples include regulation of electric utilities by the many local public utility commissions, the leasing of state and federally owned property for energy resource exploitation, regulation of pipelines by the Interstate Commerce Commission, and federal control of ceiling prices for oil (largely during the 1970s), depletion allowances, import quotas, antitrust regulations, and entitlements. Although these government activities are what we call *indirect*, they have had, of course, great influence on both the structure and the behavior of energy markets. Some of these activities have conformed to our definition of the government's appropriate role, in that they increased the efficiency of the market while dealing with such public policy issues as equity, transfer payments, and national defense. Still others decreased the efficiency of the market, and even failed to serve well their intended goals of equity.

We certainly would not argue that wherever the market leads is "good." Neither would we argue that all that is needed is a good national energy policy. What we do argue—and a thrust of our thesis—is that, in the present system,

- Decisions in both the private and public sectors must be made with a recognition of the real situation and supported by reliable data.
- The fundamentally differing, although complementary, roles of the private and public sectors must be appreciated.

We do not intend to attack or defend either the market or government policy. We believe that poor analysis and poor decisions occurred in both areas. The proper functioning of the systems as they actually exist requires the best possible analysis and decisions in both areas, and the proper differentiation between the two.

Our analysis follows the premise that blunders occurred in three dimensions: inappropriate decisions in the private sector, using outdated conceptual frameworks; inappropriate decisions in the public sector, again using outdated conceptual frameworks; and confusion about the appropriate roles of the public and private sectors and the relationship between the two.

The lessons we have drawn consist largely of conceptual frameworks, decision criteria, and contexts that are more appropriate to the conditions found in the New Energy Era. The prospects for the future presume that these lessons will be exploited, that new institutional forms will arise, and that systematic approaches will deal effectively with the kinds of issues we have largely been blundering through until now.

Energy issues must be dealt with systemically. Energy systems must be clean, safe, economically viable, robust, and technologically available. This approach is called Integrated Energy Systems. A humbling postscript is that the United States is at this point largely lagging in the effort. For example, one can see significant progress and even concrete examples of serious planning and implementation of integrated systems in Northeast Asia, Scandinavia, and Eastern Europe.

Experts are not always right. They may not even usually be right, when in times of uncertainty and crisis they are desperately needed. Perhaps, at most, experts can be helpful simply because they are the best available, if care is taken to recognize them for what they are.

We count ourselves among the energy experts who have contributed to the problem. As energy experts, we can become more helpful in the future by learning from our mistakes.

Churchill commented about a colleague, "He is humble. He has much about which to be humble." To our chagrin, we energy experts have created for ourselves a rich resource of mistakes about which to be humble—and from which to learn. The challenge we now pick up in this volume is to exploit this resource.

The book charts the history of the U.S. energy sector over the last forty years, investigating how and why mistakes were made, and how to prevent such blunders in the future. The emphasis is on the analytical framework underlying decision making in the inherently uncertain energy sector. By understanding what was wrong with the analysis, we can better understand how future decisions should be made. We can understand how analysis should be performed, how our models should be built and used, how better to connect model building with the decision-making process, and what kinds of solutions should be sought.

Part I deals with blunders. Chapter 1, "The Energy Context," presents a history of the blunders: the actors involved, the types of decisions made, the mistakes made, and how the significance of the mistakes has varied over time. Although mistakes were made in the period prior to 1973, for the most part they had little perceptible impact at the time. After 1973, both the number of major decisions required and their downside potential increased. A number of major mistakes that had long-term and far-reaching impact as well as immediate consequences for the energy sector were made. Currently we have a breathing space in which to reflect on our mistakes.

Chapter 2, "Changes in the Energy Situation: 1945–1988," describes the changing energy situation in the post–World War II era. We identify three separate energy periods, each characterized by a different set of demand, supply, and price relationships, and by a different set of issues that resulted. This chapter outlines, in broad terms, the policy followed and the underlying analysis upon which the policy was based. It demonstrates that policy actions taken during the second period compounded blunders made during the first.

Our initial assessment of the situation in 1973–1974 was almost completely wrong, but *the root of the problem was far deeper than expected.* At root was that most decision makers never really understood the degree of uncertainty inherent in the energy sector, or what that meant for successful management by government or corporations. They believed in the accuracy of their models without giving due attention to the assumptions on which they were based or to the simplicity of the optimization criteria.

Chapters 3 and 4 catalog the specific blunders that were made, with special attention to the critical second energy phase, 1973–1982. Here we draw liberally on the unfortunately rich supply of examples.

Chapter 3 describes the U.S. government's assessment of the changing energy situation, the relatively passive role it played during the first energy period, and the popularly supported decision to intervene with a heavy hand at the start of the second energy period. By way of illustration, we detail governmental policy on pricing, conservation, synfuels, photovoltaics, renewables, gas, and nuclear energy—outlining the blunders made in each area during the second energy period. This chapter also describes the gradual withdrawal of government from active participation in the energy sector during the third energy period.

Chapter 4 is a parallel discussion, describing industry's mistakes during the three energy periods: industry's analysis, its aims, and its decision-making framework. We single out for detailed attention the electric utility industry and the industry supplying it with technology and equipment. The chapter describes the kinds of questions faced by the utility industry, such as the primary energy mix, the mix of conversion technologies, the size of plants, and the timing of additions to capacity. We discuss issues relevant to these decisions, such as questions of meeting demands, efficiency, capital costs, prices, reliability, safety, and environmental damage. We detail the blunders made in decisions on nuclear plants (e.g., poor reliability, high costs, the record of cancellations, and the need for retrofitting plants under construction), and their more far-reaching impacts, the loss of leadership in technology and the virtual stagnation of the entire U.S. nuclear industry.

Part II deals with lessons. Chapter 5 describes the many reasons for the blunders described in the previous chapters. The reasons vary from an almost blind belief in analytical models with simplistic assumptions to the misjudgment of societal values.

Chapter 6 examines the purpose of modeling, and appropriate ways to do it. It considers the sophisticated assumptions embedded in simple modular models, why resource depletion is not the issue, and examines the dynamics of technological evolution. Finally, we emphasize the importance of understanding the different roles science and engineering play in most of our societal problems. Overemphasizing science and underemphasizing engineering is one of our fundamental problems.

The blunders detailed in the early chapters set the stage for the lessons of Chapter 6. The earlier chapters point toward the characteristics that will be required for the energy systems of the next twenty-five years. If the lessons learned are obvious once they are pointed out, so are the prospects. The energy systems required are, a priori, neither more complex nor simpler; they are different. Flexibility, reliability, security of supply, robustness and adaptability to new technologies, and environmental cleanliness will define the systems needed to get us into the twenty-first century. To achieve those objectives and maintain economic viability, the systems must take advantage of greater potential for integration. Much as a steel mill or an oil refinery is an integrated system, so will it be necessary for more of our energy and environmental systems to be fully integrated.

Part III suggests prospects for the future. Chapter 7 describes what we believe will be the structure of energy systems in the twenty-first century. They are referred to as Integrated Energy Systems. On the surface they seem to defy the laws of current economic and business practice. They are horizontally rather than vertically integrated. They involve multiple inputs and multiple outputs and, frequently, some type of intermediate product. The systems do not defy economics; to the contrary, they improve it.

Chapter 8 emphasizes the need for flexibility in adapting energy systems to suit the specific needs of each nation, region, and culture. It also details the practical implications of continuing uncertainty in the energy sector.

Two of the three authors have spent most of their working lives in corporate America. All three of us sincerely respect it. We do not subscribe to or even remotely suggest any conspiratorial hypotheses. We have been close enough to the world's energy leaders to be sure of this. There are no culprits to identify and punish. There are no conspiracies or conspirators operating to make immoral profits by oppressing the people. In fact, the opposite is largely true. Many of these leaders and experts have lost the most. They have lost personal and corporate fortunes in blind and vain pursuit of erroneous beliefs. How and why are the erroneous policies and actions still going on? Why have so many been so wrong for so long?

We, the authors, have come together in many different contexts, but the context in which our task was conceived was the involvement in the International Integrated Energy Systems Consortium (IIESC). The consortium is composed of six organizations in Asia, North America, and Europe, and is coordinated at the Massachusetts Institute of Technology (MIT) in Cambridge, Massachusetts. The authors are its cofounding director,[1] the director, and the principal investigator. We are grateful to our colleagues in the consortium for their encouragement.

Our book has a distinctly U.S. perspective, simply because that is our perspective. We have not tried to disguise the fact. But our experience, interest, and concerns are international. We earnestly believe that the blunders we describe are not confined by national boundaries, and we trust that our analysis is not limited in its value and applicability to a single country.

Each author brings to this work between twenty and forty years of experience in different phases of the energy sector: petroleum, nuclear, electric generation and transmission, conversion systems, and exotic technologies. In addition, each brings experience from the top levels of industry, academia, and consulting. And our experience was gained on six continents.

Although we have worked with governments in many different roles, none claims to be highly experienced in the formulation of public policy. Our perspective of public policy is largely from the private sector. The reader should understand that this perspective is as valid as the one gained from the public sector, that is, from inside the government. In addition, the perspective presented here is a critical one, for the major power the government has over energy in the United States is its power to influence—set the rules, boundaries, and context for—private-sector supply, investment, and demand decisions. These are the decisions made by the purchasers and sellers of energy in the marketplace. Decision makers range from homeowners and automobile drivers to gasoline station operators and plant managers, to the great oil companies and the captains of industry and commerce. In the United States, these decisions *are* the energy market.

## NOTE

1. With Dr. Wolf Hafele, formerly of the International Institute for Applied Systems Analysis (IIASA), currently director general, Kernforschungsanlage Juelich GmbH, Juelich, West Germany. IIASA is an international research institution located in Laxenburg (Vienna), Austria. Founded in 1972, it is supported by scientific and research organizations in sixteen countries—both market and planned economies.

## REFERENCE

Hegel, Georg Wilhelm Friedrich. 1932. Introduction to *Philosophy of History*. Quoted in *Familiar Quotations*. 1955. Edited by John Bartlett. Boston: Little, Brown, p. 401.

# Part I
# Blunders

# 1 The Energy Context

Short of a war or perhaps the Great Depression, one would be hard put to think of a single phenomenon with a more far-reaching impact than the energy crisis. It is now over fifteen years old, and, appearances to the contrary, it is with us still.

Although expenditures for energy account for only 5% to 10% of the U.S. gross national product (GNP), energy is the engine that drives industrial society. The replacement of human and animal labor with commercial energy through the use of machines is basic to the very process of industrialization. Without machines using commercial energy, industrial society would literally regress 250 years, to *pre*-industrialization. When energy is threatened, draconian scenarios are inescapable. Many authorities observe that the Japanese war against the United States became inevitable once the United States had severed the Japanese oil supply in the summer of 1941; the same authorities are only amazed that it took months rather than days for the Japanese to react. Although the underlying causes of the war were many and complex, the disruption of Japan's oil supply was certainly one of the major factors.

Although its visibility waxes and wanes, and its manifestation undergoes constant metamorphosis, the energy crisis has not gone away. Most of us suspect that sooner or later, the energy discontinuities and uncertainties will raise their ugly heads once more, to shake our world in some unpredicted way. The consequences will be unpleasant for large segments of society—exactly which segments, we cannot be sure.

The United States was one of the two targets of the Arab embargo that began that fateful thirteenth day in October 1973. Most can recall our early responses to the event. The fact that we assumed it really was an effective embargo may well have contributed to our gross misinterpretations of the event, and to our invalid responses. This assumption led us to further misinterpretations and bad analyses and still other false assumptions.

If it really was an embargo, then it was a political issue, and the solution was getting access to alternative supplies. This was the position taken by the U.S. government, a position that would find its official expression in Project Independence. Emphasis was placed on increasing the domestic long-term supply of primary energy sources. These would fill the gap, which it was thought the embargo had created, between supply and demand. U.S. coal was abundant, but it was dirty. Nevertheless, it soon came to be considered a big part of the solution—*because we had so much of it.* In 1977, President Carter announced that by 1985, the United States would be burning more than a billion tons of coal per year. This would have been a 67% increase over then-current levels (National Energy Plan, 1977). Since it was dirty, we would have to bear the economic burden of cleaning it. On the other hand, clean-burning gas was viewed as too valuable to use in utility and industrial boilers. Therefore, we prohibited such usage. "The demand for gas is considerably higher than the amount that can be supplied," we were told. Hence, gas is still rationed by prohibitions on hookups for new homes in some areas.

## THE U.S. GOVERNMENTAL VIEW

### Synfuels

One answer to filling the gap was to manufacture liquid and gaseous hydrocarbon fuels from more plentiful solid domestic resources such as coal, oil shales, and biomass. These products are generally referred to as synfuels. Admittedly, they would be expensive, but the price of energy was perceived to be spiraling ever upward. It would be only a matter of time, we thought, before synfuels would become economically competitive. Furthermore, since the price of synfuels was assumed to be stable, synfuels would set a ceiling on world oil prices. Certainly, based upon assumed continuing oil-price escalation, they would be economical within a few years, or within a decade or so at the most. In any event, they were needed as soon as possible to make the United States independent of petroleum imports. The United States was already importing half of its petroleum needs, and demand would clearly increase, since economic growth had always required a proportional growth in energy usage.

The need for synfuels seemed equally clear when viewed from a global perspective. Every schoolchild knows that petroleum is a limited and nonrenewable resource—"They aren't making it anymore." The world must eventually run out. We even went so far as to assume oil-exporting nations were saving valuable petroleum resources for future generations. We went still further by asking, "Shouldn't we be preserving ours for our future generations, too?"

It seemed clear that the entire world faced an inevitable gap between increasing oil demand and decreasing supply. At some point, demand would actually exceed supply, and the world would actually *need* synfuels. Impressive looking and scholarly appearing studies actually calculated the magnitude of the gap.

Many technologies were available to produce synfuels, some of them quite old. But since they were not economical, the question facing the U.S. government was how to get them into production. The government decided to stimulate the construction of commercial-size demonstration plants. The move was expected to initiate a decline in the cost of the synfuels, which would soon equal and eventually be less than the ever-rising prices of petroleum. When that happened, synfuels would be a viable industry, and would solve the energy crisis.

### Renewable Energy

The official U.S. position was that the earth was running out of oil. "The principal oil-exporting countries will not be able to satisfy all the increases in demand expected to occur in the U.S. and other countries throughout the 1980s. The world's presently estimated recoverable oil resources, at a conjectural growth rate of 5 percent, would be exhausted by 2010." [1]

Since fossil fuels are a limited resource, a natural conclusion was to replace them with renewable sources, which would *never* run out: biomass, solar, wind, hydro, ocean thermal, tidal, and geothermal. Renewable sources had—or appeared to have—the additional advantage of being clean, in contrast with dirty, polluting fossil fuels. Some of the technologies involved were relatively new and exotic, such as photovoltaics and fuel cells. Nevertheless, they all suffered from the same economic affliction as synfuels: they were not cost competitive. Even the newer technologies offered little hope of adequate cost reductions if realistic commercial standards were applied. Therefore, the U.S. government applied the same prescription as it used for synfuels: it would motivate the construction of commercial demonstrations that would enable a new synfuels industry to eventually compete on its own with fossil fuels.

### Conservation

A later step in the solution embraced conservation, that is, the reduction in energy usage *for the sake of reduction itself*. This popular concept often went beyond economics into the realm of human virtue.

President Carter told us that the energy crisis was the "moral equivalent of war." In a war, one makes sacrifices without regard to the economic cost. We were told to heat living spaces to temperatures below the comfort level, to insulate beyond the economical point. The U.S. tax code was amended to provide incentives for taking conservation steps beyond the point of justifiable economics. The ethic of conservation for its own sake was aimed at reducing demand, thus reducing the perceived "gap" between supply and demand.

Closely related to this philosophy was an underlying belief that we were going to have to change our national life style. We were using too much energy. At the extreme end of this belief was a back-to-nature philosophy, an extension of the environmental ethic that had gripped segments of the American public in the late 1960s and was in full stride when the energy crisis began. A more moderate and popular view was the so-called soft energy path, which rejected the conventional centralized generation and use of energy.

Thus independence from petroleum imports became—and in many circles remains today—central to a solution of the energy crisis. The goal of independence began as a strategic response to U.S. dependency on supplies of imported (read "OPEC" or "Arab") oil, which was considered an unreliable source.[2] The supply disruptions of 1973–1974 were taken as clear evidence of the problem. The supply disruptions of 1978–1979 confirmed that view.

Independence from petroleum imports would be needed eventually, since common wisdom said Americans were running out of oil. And, at some point, the whole world would have to switch from petroleum, too. Ensuring *access* to what limited supply was left became an urgent priority. Concern among importers and consumers over obtaining access to the existing supply resulted in their bidding prices up threefold. And the higher price level seemed to be only a base for further escalation, strengthening arguments for energy independence. Take-or-pay contracts for oil, gas, or liquefied natural gas (LNG) were enthusiastically pursued by purchasers; almost no terms seemed too harsh to the purchaser, if only an ensured supply could be obtained. There wasn't going to be enough to go around, so it was every man for himself.

### Mistakes

What were our mistakes? Their names were legion. We misinterpreted events, made inappropriate assumptions and analyses, misused the laws of economics and commerce, and failed to use our understanding of the relative roles of the public and private sectors.

The oil embargo was not an effective embargo at all. It was a short-term political sanction by Saudi Arabia against the United

States as punishment for its aid to Israel during the 1973 Arab-Israeli war. It could not be effective because of the efficiency[3] and fragmentation of the world tanker market. It was no more possible for OPEC to keep its oil out of U.S. supply lines than it was for the United States to keep its embargoed grain out of Soviet silos several years later. The embargo was circumvented by simply rerouting through the international system. The significance of the embargo lay in its symbolism. It marked the takeover of production levels by the host countries, serving as a clear symbol of their newly acquired power. They now were able to dictate to the international oil companies not only how much oil they could produce but also *to whom they could sell it, and under what conditions.*

The process that underlay the OPEC takeover had been going on for years, but the overt takeover and the public perception of its consequences were sudden and dramatic. The oil-importing nations were convinced that OPEC could constrain supply, and this belief permitted their fivefold price increase to hold for years without question. Availability of supply was never the problem; a discontinuous price increase was the problem.

Energy independence for the United States has never been either economically reasonable or even technically possible. Our *projections* of increasing demand and decreasing supply surely resulted in a gap. In 1977, one respected authority actually predicted that by the year 2000, demand would exceed supply by twenty million barrels per day (Wilson, 1977, Figure 8-8, p. 250). *Any* gap between demand and supply is a logical absurdity. In practice, it is simply impossible for demand to exceed supply—for us to use more than we produce. Our projections ignored the price elasticities of both demand and supply, as well as the dynamics of petroleum exploration and reserve additions. As oil prices increase, it makes good economic sense to spend money to increase conversion and end-use efficiencies; it makes good economic sense to spend money to cogenerate, switch to other fuels, and take a host of other actions, each economically evaluated on its own merits and resulting in lowered demand. Similarly, it makes good economic sense to increase petroleum supplies: increase exploration, drill deeper and in more hostile environments and in less attractive locations.

Our mistakes relative to gas were as great as those relative to oil. We were told that gas was intrinsically very valuable, and its use was regulated accordingly. But gas has no intrinsic value. Its only value is economic and can be so appraised.

The mistakes relative to coal were equally great. We were urged to use it because the United States possessed vast reserves of it. "We have more coal than Saudi Arabia has oil." But to use more coal because we have so much makes no more sense than it does to recover iron from seawater because there is so much seawater.

Add to these missteps those of Washington. Although the U.S. government is a large landlord of mineral rights, it is neither a producer nor a consumer of coal—nor any other form of energy—to a significant degree. Its primary power is its ability to indirectly influence myriad private decisions by dull policy instruments, such as taxes and subsidies. Further, its role in the energy sector is limited. But that did not stop it from trying to play a major part. The same is true of its role in the commercialization of new technologies, even though commercial-scale demonstrations of intrinsically uneconomical technologies are unlikely to accelerate their commercialization. The direct replacement of petroleum, which is in short supply, with a synthetic material that approximates its qualities (i.e., synfuels made from coal or natural gas) is irrational until economics dictates the substitution. Renewable energy sources, as well as synthetic fuels, have no intrinsic value; they become attractive only as they become cost competitive.

One might argue that there is a time lag between rational market-oriented decisions and rational social decisions. To the extent a time lag actually exists, the government has a legitimate role to play.

Energy usage is not inexorably linked to economic growth in a predetermined relationship. In the long term, natural economic forces will moderate rising energy usage and prices by a rational rearrangement of the proportions of capital, energy, and labor. In mistaking short-term events for long-term trends, we did not allow for the time required for rational economic responses. We assumed that the period 1973 to 1979 was a harbinger of the long term, when in energy dynamics it was in fact the short term. It seemed that the laws of economics had been suspended. We therefore jumped to the conclusion that reductions in energy usage must be imposed, or that they would come from declines in economic activity, or that they would result from partial deindustrialization.

## THE U.S. PRIVATE VIEW

### The Electric Utility Industry

The views, actions, and mistakes of the private sector complemented those of the public and the government. In stark contrast to the earlier-mentioned soft energy path was a hope—rapidly fading as it turned out—that nuclear power would provide a significant part of the solution. If not providing an all-electric society, nuclear energy might at least replace fossil fuels as a source of electricity, not only reducing dependency on fossil fuels but also reducing pollution and costs. Not surprisingly, this view was promoted by the electric utility industry, which foresaw ever-increasing electric demands and a need for new plants. Given what was then understood about efficiencies and economies of scale, the plants escalated in size. This escalation was carried out without adequate attention to reliability and, hence, real operational economics.

It is now well known that in placing our hopes on nuclear energy, we overestimated the technological and economic qualities of the option, while undervaluing the seriousness of the public concern over safety and long-term nuclear waste issues.

### The Petroleum Industry

The petroleum industry was as much in sync with the public as was the electric utility industry. But the public was not entirely in sync with the petroleum industry, and the public attitude toward it was—and remains—at best schizophrenic. On the one hand, the public supported the industry's exploration and production activities. On the other, the government endeavored to control the prices of domestic oil and limit what came to be known as windfall profits through an inscrutable web of price controls, taxes, and regulations (e.g., crude entitlements and the small-refinery bias).

Believing that the nation and the world were running out of petroleum, the industry set out to discover and produce as much as possible. Exploration efforts found funds without difficulty, and almost without question, as operators and investors alike bought the premise that prices were on an unstoppable upward spiral. Not surprisingly, discoveries were made that could only be economically viable at elevated prices.

As the custodian of petroleum, the petroleum industry became the self-appointed, but unquestioned, custodian of synfuels, most of which were, in effect, synthetic petroleum. A mutual admiration society was formed by government, the petroleum industry, and that segment of the public which favored—for very different reasons—the synfuels solution. The petroleum industry was to give birth to the new industry, with government as midwife and the public as cobeneficiary if not cheerleader.

The underlying mistakes were many. We are not running out of oil, and do not automatically need a synthetic replacement. Even if a synfuels replacement were appropriate, the petroleum industry might not have been its most fitting parent or custodian. And, of course, prices can never spiral only upward as we had assumed.

## The Consumer

In the story of the emperor's new clothes, one factor that disguised the ruler's nakedness was mistaken groupthink among the onlookers. So, also, in the energy crisis. Underlying the high level of confusion, disagreement, and debate was an amazing consensus among sectors on crucial but mistaken assumptions. The consensus made the mistakes much harder to see for what they were.

The consumer joined government and industry in accepting the assumptions that we were running out of oil, that prices would spiral ever upward, that we needed to conserve for virtue's sake, that economic growth marched lockstep with energy use, and that government needed to see to it that we got the synfuels and renewable energy sources we needed.

At the same time, new public concerns were entering the national consciousness. They were to have a profound impact on the way we perceived and responded to the energy crisis. One was concern over the environment. What started as the minor cause of a few scientific visionaries and social activists grew, within a couple of decades, into a global consensus that industrial progress at the expense of the ecology was simply unacceptable. It was during the energy crisis of the 1970s that the concern gained its widest and most effective support, and when its revolutionary ideas were institutionalized in legal form and government agencies. We did not, however, take the environmental impulses of the public fully into account in responding to the energy crisis because of two related factors. First, industry largely assumed that cleaning up the environment was a necessary evil, an added cost to be minimized, avoided, or delayed. This assumption has turned out not to be necessarily the case. We are now finding that less costly energy systems can be more environmentally benign.

Second, and at least as important, many who were devoted to dealing with energy issues did not appreciate the linkage between environmental and energy issues. Energy people and environmentalists formed opposing camps. Either we would have enough energy or we would have a clean environment.

The fact is, many environmental issues are directly related to energy in one of two ways. One is the conversion or processing of energy from one form to another (e.g., the refining of crude oil into petroleum products such as gasoline, or the generation of electricity from fossil fuels). The second is the final use of energy (e.g., coal in furnaces to make iron and steel from ore, gasoline in automobiles or kerosene in airplanes to provide transportation, natural gas or heating oil in domestic furnaces to provide home heat, or residual fuel or coal in industrial boilers to generate plant steam). The two forms of energy-related processes are among the greatest contributors to air pollution. Therefore, if the environment is to be cleaned significantly and if pollution is to be reduced drastically, a major portion of the task must be accomplished by somehow altering the processes that convert and utilize energy. Energy stands at the center of the air pollution issue.

Much of the public at large has come to see nuclear energy—from fuel preparation through heat generation to waste disposal—in a special light. Whereas pollution from fossil fuels occurs in undramatic ways, that from nuclear is dramatic. Industry and government failed to take the difference into account when pursuing the nuclear option. In evaluating pollution and safety of nuclear energy, the public was unwilling to consider the issues from a statistical or levelized perspective. For example, statistical analyses concerning lost time or deaths per million man-hours from oil field accidents compared to those from nuclear plants simply carry no weight. The absurdity of the following statement, made in all seriousness, will illustrate: "A nuclear power plant is infinitely safer than eating, because 300 people choke to death on food every year" (Cerf and Navasky, 1984, p. 216).[4] Whether we like it or not, the specter of a nuclear catastrophe carries a logic of its own.

## BLUNDERS, LESSONS, AND PROSPECTS

Just as the term *embargo* misled us, so has the term *crisis*. In common parlance, a crisis is a relatively short-term phenomenon from which one either dies or recovers. We now know that the energy phenomenon does not fit this description. Thinking that it was a short-term aberration, which raised problems to be solved, we missed valid approaches and considerations. What is the real nature of the seventeen-year-old phenomenon, from which we have neither died nor recovered, which has not gone away? What were our real blunders? Why did we make them? What can we learn from them? And what are our prospects for the future? These are serious questions that deserve serious answers. It is to that task that we now turn.

## NOTES

1. National Energy Plan. 1977. Washington, DC: Office of the President, Energy Policy and Planning.

2. The goal of independence had its roots in the Import Quota Program, which started in the late 1950s. The stated purpose of the program

was to reduce U.S. dependency on oil imported by water, since such a source was considered unreliable in the event of a war. The effect of the Import Quota Program will be described in more detail in Chapter 2.

3. The term is used in its technical economic sense.

4. Cerf and Navasky are citing Dixy Lee Ray, governor of Washington and former chairman of the Atomic Energy Commission, 1977.

## REFERENCES

Cerf, Christopher, and Victor Navasky. 1984. *The Experts Speak.* New York: Pantheon, p. 216. Quoted from Botts, *Loose Talk*, p. 7.

National Energy Plan. 1977. Washington, DC: Office of the President, Energy Policy and Planning.

Wilson, Carroll L. 1977. *Energy: Global Prospects 1985–2000.* New York: McGraw-Hill.

# 2 Changes in the Energy Situation: 1945–1989

In the post–World War II period there were three very different energy phases. The first, from 1945 to 1973, was a long phase of stability when energy prices were low and supplies were stable and secure. Energy was not a topic of great interest outside the industry itself.

In the first phase, low petroleum prices dominated the energy sector worldwide. The low price naturally resulted in petroleum gaining market share in primary energy markets, and in keeping low the prices of other, competing energy forms. Investment decisions in energy conversion and usage established in the industrialized society a strong dependency on high energy usage in general and especially on petroleum. The sheer magnitude of these conversion and usage investments ensured that the dependencies were fixed long into the future.

Then energy catapulted to center stage. Prices of all forms of energy jumped dramatically as the oil-exporting nations took control of oil production rates. First in 1973 and again in 1978, oil supply to the industrialized world was disrupted.

During the second phase, from 1973 to 1980, complacency over energy was replaced by a crisis mentality. There was the shock of escalating oil and other energy prices, the realization of the extent of our energy dependency and of the political consequences of relying on foreign sources of supply, and innumerable forecasts from usually reliable sources of impending resource exhaustion. All these factors contributed to decisions to invest in new and renewable forms of energy, in programs for energy independence, and in accelerated nuclear power development beyond what was later to be known to be a reasonable pace. Most of the ensuing research and investment failed to achieve a significant shift to new, nonpetroleum fuels. And the underlying notion of energy independence is now recognized as a false hope.

Until 1949, the United States had been a net exporter of oil. But U.S. demand continued to outstrip domestic supply; by 1973, the United States was importing over one-third of its oil demand. Since then, despite periods of the highest imported oil prices in history, and despite all efforts to reduce demand and increase domestic supply, imports continue to account for between a third and a half of U.S. demand. It is not feasible for the United States to even aspire to be independent of energy imports, given the technological society and global economy into which we have evolved.

Finally, in 1981, we entered what appears to be a different phase—one likely to be temporary. The world has a large surplus capacity for primary energy production and for electricity generation. Once again, primary energy is cheaper and more abundant and supplies are, apparently, relatively secure. But this phase is very different from the first one described above. It is the result of the high prices of the second phase and the overreaction to them. Hence, it must not be mistaken for a period of stability.

In the third phase, total uncertainty has replaced the presumed stability of the first phase and the repeated upward price discontinuities of the second. So far during the third phase, prices first eroded, then collapsed, and then partially recovered—only to erode, collapse, and recover again. These occurrences, if misinterpreted, could be highly dangerous, lulling us into a false sense of security. On the other hand, if we use the time we have prudently, the third phase may give us a second chance: to reflect on our mistakes, learn from them, and avoid their repetition. If we choose that course, energy prospects could become quite bright indeed.

We begin our analysis by charting, in more detail, the changing energy situation from 1945 to the present. For each phase we describe prevailing conditions of supply and demand and the issues and concerns to which they gave rise. We also consider how energy analysts contemporaneously viewed conditions and why, so often, their understanding of key relationships and factors was wrong. Finally, we look at the decisions that were based on misconceptions and place them in the context of changing societal values and pressures. Only in the context of a changing value system can we see how wrong the advice of our energy experts most likely, and indeed, was.

## THE FIRST PHASE: 1945–1973

For our purposes, the crucial features of the first phase were that prices of energy were low, and prices were stable. The prices of all forms of primary energy (coal, natural gas, hydroelectric, and so forth) were held low by the abundance of low-priced oil. Most of the oil traded in the world market was very low-cost oil from a handful of third world, largely Middle Eastern, countries. But it was controlled by private, Western-based, multinational corporations—including the so-called Seven Sisters.[1] Five factors shaped the global energy sector during this phase. Of these, the most central is that in the free world over the last forty years, crude production capacity has always exceeded demand.

The second factor is politico-economic. The Western companies that dominated the industry—at least in the first phase—were constrained at least in some measure by an antitrust environment, for most of the international oil companies were domiciled in the United States, where the antitrust restrictions have traditionally been the most stringent.

The importance of this factor is that although the Western international oil companies controlled the levels of production of crude oil entering the world markets, they were constrained in allocating production among themselves. It is certainly true that in many important areas of the world the companies produced crude literally in partnership with one another: for example, Chevron, Exxon, Mobil, and Texaco through the Arabian American Oil Company (Aramco) in Saudi Arabia; British Petroleum and Gulf through the Kuwait Oil Company (KOC) in Kuwait; and all Seven Sisters through the National Iranian Oil Company (NIOC) in Iran. Within each arrangement, some effective production allocation patently occurred. On a global basis, however, this was not the dominant factor in controlling surplus production capacity. (As we shall see shortly, the dominant factor was the setting of production levels, not by producers, but according to the various companies' *ability to market* the oil.)

Of course, there is a popular argument that by controlling production, the industry thereby controlled prices—presumably at a level higher than they would otherwise have been. Whatever may be the merits of the argument, the important point is that, compared to the entire period following 1973, prices during the first phase were both *low* and *stable*, and supply was *dependable*. We are interested in what happened to change low and stable prices into higher and unstable prices, and change supply from dependable to undependable.

The remaining factors are more purely economic. The third is that petroleum was increasing its share of the energy market at the expense of coal. For this to continue, the price of oil would have to remain significantly below that of coal (on an equivalent $/BTU basis).

The fourth factor is that oil is a classic commodity: oil of a given grade produced by one company or country is indistinguishable from that produced by another. Given adequate refinery configurations—which in fact exist—one grade can be satisfactorily substituted for another at a differential value, which can be technologically determined. For example, the difference in value between a light and heavy crude, or between a sweet, low-sulfur crude and a sour, high-sulfur crude can be determined with technological accuracy. There is little, if any, opportunity for product differentiation, and worldwide transportation is cheap. Therefore, oil was bought and sold in an efficient (using the word in its technical economic sense), worldwide market. The last factor is that very high fixed costs (discovery costs) and relatively low operating costs (production costs) are endemic to the exploration and production business. Oil is risky to explore for, expensive to discover, but cheap to produce.

The result of all five factors was a low and stable world oil price, with no perceived shortages or supply disruptions. The price was low compared to what followed in the wake of OPEC's control of the very same production capacity. Until then, the oil companies were eagerly competing with one another, with adequate supply and all types of commercial favors (e.g., loans at attractive terms for refineries or tankers built in the purchaser's country gratis, and free technical assistance and training for the purchaser) for every potential purchaser of crude oil.

Together these ingredients would seem to be a prescription for disastrous price wars and instability. But the controlling companies achieved de facto stability for the entire global energy sector by means of vertical integration. Instead of price wars at the wellhead—a game OPEC members eventually played against each other beginning in the early 1980s—each of the large producers developed its own markets for its own crude by building and operating its own transportation, refining, distribution, and marketing networks. Thus, the battles were over customers for products, rather than over wellhead prices and a concomitant flooding of the world with production that exceeded demand. Although production *capacity* exceeded demand, actual production did not. Final product demand, not production capacity, determined production levels for each major producer, and for the industry as a whole.

The operation of the global system was showcased by that of the Texas Railroad Commission, which set allowable production rates for all crude production in Texas. In 1973, Texas still accounted for well over one-third of U.S. production. (Even today, Texas remains the largest U.S. producer-state, including Alaska.) Formal legal structures similar to the Texas Railroad Commission were established in other large producing states, such as the Louisiana Conservation Commission. This type of arrangement formally dominated U.S. production. The pattern was also followed more informally worldwide throughout the first phase.

A key to the success of the operation was that production levels were set, not by producers, but by marketers. For example, it was the *refiners* who appeared before the Railroad Commission, to nominate the quantity of crude they would *purchase* during the ensuing month. Admittedly, in most cases, the producers and the refiners-marketers were the same companies, that is, Exxon, Mobil, Texaco, and so forth. But the point is that the balance of power did not lie in the hands of independent producers, who had neither refineries nor markets.

Herein lies the crucial difference between phase one and those that followed. In the latter two phases, production levels are set entirely by the independent producers, the exporting countries, who on the whole do not control significant refining capacity or significant markets. It is interesting to note that these exporters are beginning to integrate downstream by purchasing interests in refineries and markets traditionally controlled by integrated companies. The process has long been recommended by one of the authors as a means of recovering some stability in the industry and in the energy marketplace (Ball, 1985).

The result of the arrangement in phase one was not only stable prices but also what we now in retrospect recognize as relatively low prices. Though their vertical integration afforded stability, the economic laws of competition and the statutory antitrust laws prevented the vertically integrated companies from capturing a significant portion of the consumer surplus.[2] Supporting this statement is the fact that—despite admittedly favorable U.S. tax considerations such as depletion allowances and foreign tax credits—the petroleum industry has not made significantly higher returns than industry as a whole.

If the price of oil had been kept high in some way, the penetration of oil into the energy market would have been inhibited or even prevented. Instead, the low price of oil drove technology advances in the areas of petroleum production, refining, transportation, and utilization (e.g., boilers and internal combustion engines). As a result, low-cost petroleum-derived energy captured the market from coal quickly and thoroughly. Energy consumers converted old plants or engineered new ones to use petroleum. The low price and adequate supply of oil made oil a buyer's market. Price and supply were the foundations of the oil age. The world became hooked on petroleum-based energy, and established that dependency through massive investments in oil-fueled equipment and infrastructure. Without its low price, petroleum most certainly would not have penetrated the energy market to the degree it did.

## The Low-profile Energy Sector

Throughout the first phase, energy was a low-profile sector of the industrial economy. And society was generally unaware of its critical dependency on energy use—in part, because energy was so cheap in its direct use for domestic heating, fuel for cars, and so forth, and in its indirect use as a power source and raw material for production processes such as paints and plastics. Much of the demand for energy is indirect, and that makes it less visible.

Another key factor in the low level of energy consciousness was that the full costs of energy use were not borne by the user. Users of energy were never required to pay for what is now recognized as its external costs. External costs are the damage to the environment: atmospheric pollution, contamination of water courses, spills during the loading or shipment of oil, dead birds, and other ills. External costs also cover the land usage and intrusion associated with coal mining and the siting of power plants and power lines. Agricultural land and established property rights are often lost in many hydroelectric schemes, and above all, there is concern about the safety of nuclear plants. External costs are difficult to measure, since they include society's subjective assessments of the value of clean air, open countryside, and safety. Even in today's market, they go uncharged. Throughout this phase, energy users were never fully required to take account of the externalities. The results were subsidized energy use and a diminished importance of external costs ascribed to energy.

Until 1973, nothing happened to challenge the low status we had afforded energy in our economies or our thinking. We had become highly dependent on energy for the maintenance of our lives and life styles. Despite economic cycles, the closure (twice) of the Suez Canal, frequent labor unrest in the industrialized world, and political unrest in key oil-producing and -exporting countries, we never questioned the continuity of low energy prices and guaranteed supplies. And this had both immediate and longer-term consequences for energy demands and for the primary energy mix.

## Consequences of a Low-price Energy Sector

The most obvious immediate consequence of low prices and stable supply was a high level of energy use. Unlike other costs of production—labor, capital, other raw materials—energy was cheap, and where possible it was used as a substitute. Low energy prices reduced production costs and contributed to a consumer boom. Energy-using products like cars, electrical goods, and domestic appliances featured high among consumers' preferences and further boosted the demand for energy. Low energy prices resulted in low transportation costs, thereby altering the balance of locational advantage. This shift led to higher levels of transport of raw materials, finished goods, and even people. It led to high levels of ownership and use of private cars, which, in turn, allowed for a greater separation of home and workplace, and the emergence of new life styles built on personal mobility. Cheap oil smoothed the way for the alteration of the American landscape. Suburban housing, shopping malls, and the road systems that linked them to one another and to core cities were possible, in part, because of low-priced gasoline. The low price of oil-based energy compared to other forms made oil more attractive for purposes such as domestic heating and electricity generation that made less use of some of its special advantages (e.g., as the only practical energy source for automobiles and aircraft). Clearly, low and stable petroleum prices had significant long-term consequences.

Investments in equipment and infrastructure, rational at the time, afforded no flexibility with which to deal with the impending traumas. With petroleum prices so low, for example, there was little incentive for automobile manufacturers to produce efficient cars or for consumers to buy them. Compared to the cost of a new car, fuel costs were insignificant. This was true even in European countries where petroleum was heavily taxed, but it was especially true in the United States where petroleum taxes were low. As a result we entered the early 1970s with an automobile industry tooled for the manufacture of gas-guzzling behemoths. This trend represented considerable capital and implied a long dependency on large-scale oil use.

Similarly, the low cost of travel defined the accessibility of real estate. Accessibility is a function of distance and travel costs. In turn, accessibility is a major determinant of land values and urban form. Low-cost travel created a specific climate for locational investment decisions, artificially reducing the incentive to concentrate and encouraging dispersion. The resulting land values and travel cost indicators steered investment decisions. The resulting land-use patterns established our dependency on petroleum for the future by creating reliance on, especially, (auto)mobility.

Cheap petroleum fueled the demand for cars and increased the relative advantage of private motoring over the use of public transportation. This contributed to a long-term decline of public transportation to a level where it was unable to provide real mobility for the public. By the mid-1970s, most industrialized societies—and the United States in particular—were dependent on automobiles.

These examples from the transportation sector illustrate situations that had developed in all sectors of the economy. The pattern was common, if not universal. Cheap energy led to high levels of energy usage. The reliance on cheap energy was perpetuated by heavy investment in structures, products, life styles, and processes that were highly energy consumptive. The greatest commitment was to oil, since it was versatile, cheap, and readily available. By comparison, investment in alternatives was seldom attractive. Oil had largely replaced coal, not through price alone, but also through improved conversion technologies: competitive advantages in the capital cost of oil-fired facilities, versatility, cleanliness, and ease of handling and transporting. The coal industry was on the decline from nearly every perspective.

Sliding along on a rich film of petroleum, the industrialized world was simply not equipped to respond to oil price increases—especially to sharp price discontinuities—either by reducing demand or by switching to alternative primary energy sources. In the short term, the whole Western economy was oil dependent and vulnerable to oil price increases and supply disruptions.

## THE SECOND PHASE: 1973–1980

In late 1973, without effective warning,[3] the stable energy world that had existed since the end of World War II suddenly came to an end. What followed was no random hiccup but a permanent change, the end of the old and the beginning of a completely new era. The energy situation had changed, radically and permanently. Energy prices would no longer be low and stable.

The change was brought about purposefully by events in the critical oil sector of the energy market. The cause was not the Arab oil embargo, as is frequently claimed. The embargo was instituted in a different context, as a sanction by Saudi Arabia against the United States and the Netherlands, for their aid to Israel during the 1973 Arab-Israeli war. OPEC was, however, able to exploit the opportunity, and use the embargo to symbolize its new power.

The international oil transportation market (i.e., the world's fleet of oil tankers) was far too fragmented and efficient for any proclaimed embargo to have a significant effect. Although there were short-term supply disruptions, the only tangible effect of the embargo was to increase some transportation costs slightly, because of the diversions, reroutings, and transshipments necessitated. The embargo was the ceremonial signal that the host governments had arrived at a point where they could call the shots, and U.S. companies would have to implement an embargo against their own nation, including its military bases, however ineffective it may have been. OPEC could require U.S. companies to serve OPEC's foreign policy!

The real cause of the new era was the loss of control over oil production rates and prices by the international oil companies to the governments of the exporting countries. OPEC was conceived in Baghdad in 1959 by five countries whose crude oil production represented 80% of the world's total exports: Iran, Iraq, Kuwait, Saudi Arabia, and Venezuela.[4] The immediate impetus for its formation was a price reduction by the major international oil companies. For years, the organization was impotent, but it grew in numbers and strength as it struggled continually with the producing companies for higher prices, participation in the production, agreements on production allocations, conservation policies, and uniform royalty rates. Its powers increased, and the immensity of the mutual interests of OPEC members began to overcome their immense diversities. Negotiations with the companies tended toward confrontations as its demands on the companies escalated.

The conventional relationship between an oil company and the host government was a *concession*. This was, in effect, a long-term lease to the company, which the government taxed. As early as 1920, Iran had conceived the concept of *participation*, which would give the government part ownership in production. But the concept had been ignored. When Mexico exploited production participation in 1938 and Iran in 1951, they were excluded from the world oil markets. Had OPEC perceived by late 1973 that the world had become so dependent on OPEC oil that if it were to control levels of production it could turn the world oil market from a buyer's market to a seller's market? Would rising global ecological concern over atmospheric pollution make it doubly difficult to revert to coal as the fuel of choice?

All these concerns, which had been building for decades, began to move toward a climax. Events during the closing months of 1973 moved swiftly. Concomitant with the embargo, OPEC was negotiating with the international oil companies, demanding huge price increases. Unable to reach agreement, the individual national governments took control of their own production and, acting in concert, were able to restrict the supply of oil entering the world market. Given the short-term inability of the importing nations to change their oil consumption patterns, and the dearth of non-OPEC exporters, the price elasticity of oil demand *and* supply was low. The exporters capitalized on this fact. By setting much higher prices and restricting supplies to ensure that the prices would be paid on world markets, the governments of the exporting countries were able to gain vastly larger overall revenues.

> The price of oil had . . . quadrupled in just over two months. With incredible suddenness . . . all the converging trends of 1973—the movement towards participation, the shortage, the Arab-Israel war—had come together to clinch the [OPEC] cartel. (Sampson, 1975, p. 259.)

The superficial appearance—and the popular view—was that OPEC had replaced the international oil companies in setting production levels: "the whole oil cartel had apparently fallen into the hands, not of seven companies, but of eleven countries. . . . 'We just took a leaf out of our masters' book,' as one Kuwaiti put it. The Western nations now found themselves, to their bewilderment, confronted with a cartel, not of companies, but of sovereign states" (ibid.). The word in the oil country was that OPEC had moved the Texas Railroad Commission from Austin, Texas, to Vienna, where OPEC has its headquarters. The new arrangement seemed to remain effective until OPEC's members began to disregard their own allocations, about 1981. A critical difference had appeared by then: OPEC lacked the Railroad Commission's Texas Rangers and the U.S. government's Connally Hot Oil Act to enforce its allocations. From the beginning, however, in 1973, an additional and perhaps more critical but unappreciated difference existed. Even today, it remains unappreciated, and here we break with conventional wisdom. The critical point is not whether the international majors formed an effective cartel, which was replaced by an OPEC cartel. The critical point is the change in the *basis* on which production levels were set.

Sampson (1975) and Blair (1978) rightly discern that the central issue in world oil is control, that is, control of levels of production and, by implication and even more crucial, control over *surplus* production capacity. In 1973, the shift in control was not only from one group (the international majors, whether a cartel or not) to a different group (OPEC). There was in addition a structural shift.[5]

Production levels had been determined, not by the production *desires* of the integrated oil companies, but by their *ability to market the oil.* The vertically integrated international oil companies were the link between the producing countries and the users of the crude and its products. Therefore, whether wittingly or willfully, they had played a coordinating role between supply and demand. The two opposing forces had been joined de facto.

With production control passing to OPEC, the industry was no longer vertically integrated. The oil-exporting governments became in effect the "independent producers," which never before had set production levels. Unlike the previous controllers of production, these producers did not coordinate volumes with the purchasers, refiners, consumers, and users of crude. The world had become so dependent on their merchandise that they perceived no need for such links. For the first time in the history of the world oil market, control of production levels was detached and separated from consumption.

There was no amelioration between the opposing interests of producers and consumers. In addition, the new forces controlling production operated under none of the antitrust constraints that had applied to the integrated international oil companies. Oil exporters could legally and openly agree on production levels and prices—a practice that would literally have been felonious for some of their predecessor companies.

Producers and consumers now struggled against one another without precedence and without restraint, causing supply and price disruptions heretofore unknown in the oil business.

Despite the popular notion at the time and during the hysterical decade following the embargo, the laws of economics had not been repealed. Market forces were still at work; the inviolable relationships between supply, demand, and price remained intact. But the separation of forces enabled, for the first time, the fundamentally differing interests of the oil producers and oil consumers to come to the fore. By 1973, the point of leverage had changed in a critical way, and market power had shifted from access to the *market* to control of *production*. For OPEC's control of production levels to be meaningful, it had to be able to sell its oil directly to the marketplace. By 1973, it could.

The shift in power was dramatic and had equally dramatic consequences. Three decades of tranquillity ended suddenly with oil price increases, supply distortions, chaos, and uncertainty. Of course, the balance of power, at least in the short to midterm, was in the hands of the oil-exporting countries. They were able to threaten the well-being of the entire industrialized world by means of relatively modest adjustments in oil output, or even in announcements or hints of such intentions. The consuming nations had only limited capacity to reduce oil demands without incurring huge social and economic penalties. Against the backdrop of the imbalance of power, energy issues sprang from relative obscurity to dominate our lives.

### An Unconventional View

Admittedly, we are presenting a rather unconventional view of the nature of the systemic shift that occurred in the structure of the world oil market in 1973. Our aim is not to fix the blame for higher prices. We are not concerned with whether or not either the Seven Sisters (and their cousins) or OPEC is, in effect, a cartel, monopoly, or oligopoly. Rather, we prefer to focus on the underlying cause of the shift from what we now perceive to be relatively low but very stable prices to higher but highly unstable prices. Perhaps the following description will serve to clarify our view.

The very nature of the petroleum market ordains instability. In the midterm (i.e., a few years), marginal costs of crude oil are quite low (excluding discovery costs). But in the same period, demand is highly inelastic. Therefore, producers can operate profitably at very low prices, but consumers would be willing to pay very high prices. The two opposing groups struggle over who gets the benefit of the difference (i.e., the consumer-producer surplus). Therefore, price stability can *only* be the result of very close coordination between supply and demand. When small additional supplies enter the market, price will drop precipitously. On the other hand, if even relatively small supplies are withdrawn from the market, the price will spike dramatically.

Prior to 1973, coordination was possible because the production levels were determined by the companies that marketed the oil and its products. (See Figure 2-1.) For example, Company 2 scheduled its share of production in Country A and Country B (and the other countries in which it had interests) only to match the volume it could market to consumers in the marketplace. It could accomplish this by virtue of its vertical integration, from production of crude at one end of the world to retail marketing of the finished products at the other end of the world.

Although Company 2 cooperated with, for example, Company 3 in Country B, they competed in the marketplace. The competition was in the marketplace, not at the wellhead. That is a crucial point. Without any formal or conscious allocation system, production levels were established de facto to meet demands. There was no need for quotas or allowables. There was no surplus production, and no shortages. Prices were stable, and supply was assured. Supply was not constraining; therefore, prices were stable at what we would now, at least, consider a low level.

In exporting countries there was no real market at the wellhead; the exporting countries did not compete with one another to sell oil. That is, in general, exporting countries did not sell oil. Each company producing oil in an exporting country paid the country a royalty or tax, and then the *company* sold the oil to a refiner or refined it itself and sold the products.

The role of the so-called exporting countries was somewhat comparable to that of the owners of mineral rights in the United States. They collected a royalty on the production of oil. But as the owners of the oil in the ground, they had little influence in setting levels of production, and no involvement whatsoever in the marketing of the oil.

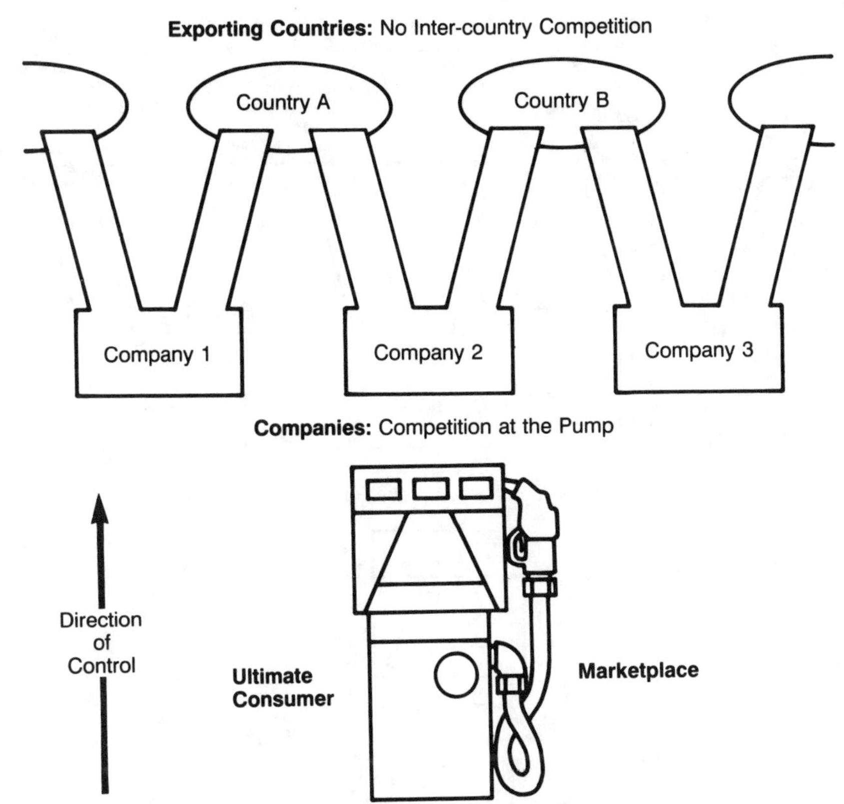


**FIGURE 2-1 Production Levels Set by Companies, Each to Meet the Demands of Its Own Customers in the Marketplace.**

In 1973, the exporting countries changed this dramatically. They took control of the level of production, and sold the oil to the companies at the wellhead. They were insulated, however, by these companies from contact with the ultimate consumer and the marketplace. (See Figure 2-2.) This would be comparable to the owners of mineral rights in the United States setting production levels to suit themselves, and then selling the oil to the companies at the wellhead.

Having lost the de facto coordination between demand and supply, price and supply became not only uncertain but also subject to violent changes. For the first half-dozen years following this structural shift, supply was constrained relative to demand, so prices jumped upward.

During the latter portion of the period, an old dynamic took on a new significance. The trading and brokering of crude in the world

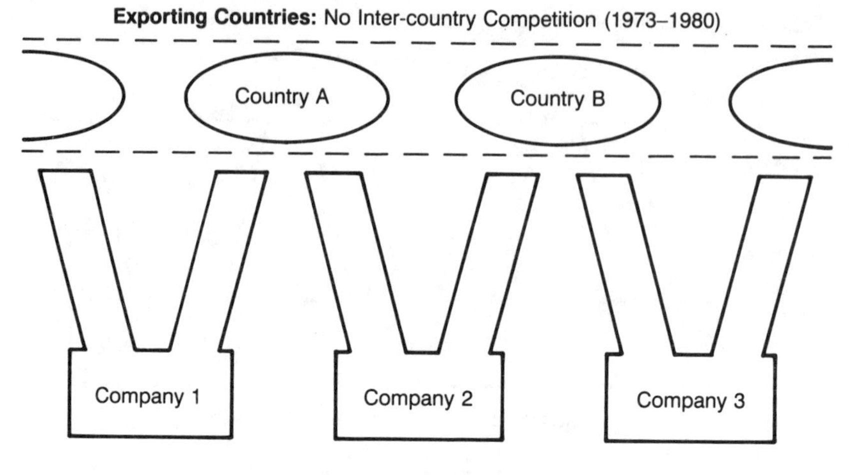


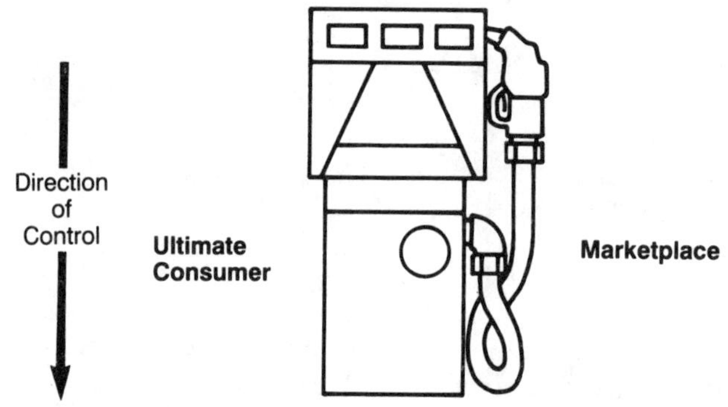


**FIGURE 2-2 Production Levels Set According to Desires, Wishes, and Propensities of Exporting Countries.**

market had always served as a balancing mechanism, and as such had not been an important factor per se. Now, however, the dynamic took on significance. (See Figure 2-3.) Now, the exporting countries began selling not only to the oil companies who had been their developers but also to brokers and traders, and even to other oil companies and to other governments. In the terms of the professional economist, the world oil market became more efficient. But it also became even more difficult for either ad hoc or institutional coordination between production levels and ultimate demand. Now, the exporting countries were in fact competing with one another in selling oil at the wellhead. They had created in essence an auction market at the wellhead for world oil.

As the desires of exporting countries resulted in unrestrained

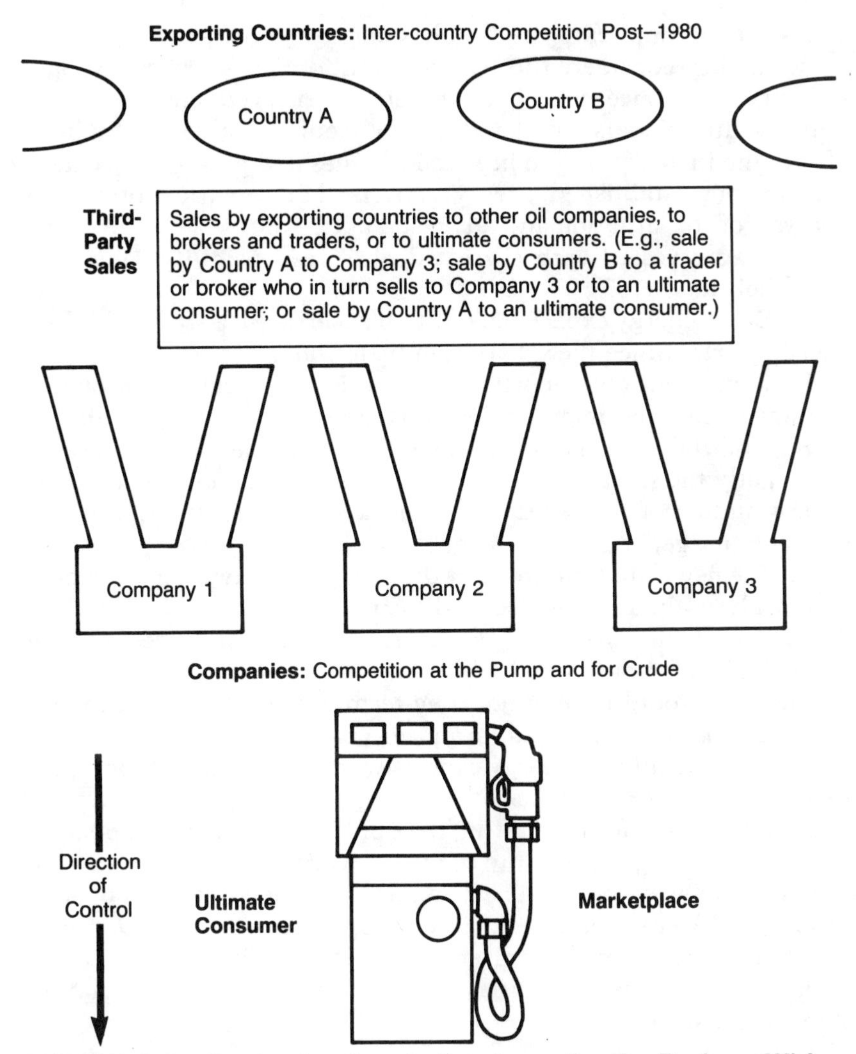


**FIGURE 2-3 Production Levels Set According to Desires, Wishes, and Propensities of Exporting Countries and/or Third-Party Sellers.**

supply, prices began to drop precipitously, but with continuing high uncertainty.[6]

## Bring in the Experts

These shifts in power imbued the industrialized world with a new-found concern for energy. The ensuing oil supply disruptions and price increases raised new issues and heightened the importance of others. Governments were concerned about the macroeconomic and strategic implications of higher energy prices and dependency on what they considered to be uncertain foreign sources of supply. Industry was concerned about the effects on production costs and markets. Individuals wondered if they would be able to obtain, or afford, the fuel required to heat their homes and power their cars. It was to long-established energy experts that attention turned for analysis of the situation and advice on policy options. Overnight the experts were vested with a great deal of responsibility. How good were their answers?

Every energy expert had built up expertise in a wholly different energy world. Since there had been little change in price over time, there had been no opportunity to study the critical economic dynamics (e.g., the interrelationships among price, supply, and demand, and the dynamics of the substitution of one fuel for another). For exactly the same reason there had been no way of developing an understanding of the relationships between energy prices, the level of energy usage, and economic growth.

The new situation was very different from anything the experts had dealt with. It was also a difficult situation to analyze. The dynamics of the systems (the relevant economic, societal, and technological factors) were highly volatile and unpredictable, yet decisions that would have major long-term consequences and involve massive commitments of public and corporate capital had to be made quickly. Above all, the energy experts were analytically ill equipped to develop forecasts and scenarios. The very techniques (econometric models) they used for forecasting were only suitable for extrapolating past and existing trends into a future where the same forces and relationships were operable. The models depended upon stable conditions and were therefore wholly inappropriate for a situation in which there were new factors, enormous changes in the values of key variables, severe discontinuities, and new types and forms of relationships to consider.

## An Abundance of Myths

Much of the initial analysis of the situation was based on a number of misconceptions: implicit assumptions that seemed self-evidently true and hence were rarely examined. These hardened into myths and, taken as conventional wisdom, formed the basis of much of the energy policy that emerged in the post–1973 period. We discuss the myths in detail in a later chapter, but it is worth stating their essence here, since they colored not only our understanding of the situations but also how we specified the problems we faced. That specification determined how we went about finding solutions.

The first myth was that the demand for energy would continue to grow. This was based on the well-known and well-understood (but wrong) assumption that energy consumption was intrinsically and *proportionally* linked to economic health and growth. It was firmly believed that for the domestic product to increase there had to be a proportional increase in energy consumption. If the industrialized world was not to go into permanent recession, and if third world countries were ever to develop at all, the demand for energy would increase in proportion to development.

The second myth was that energy prices, especially the price of oil, would continue to increase until a backstop technology (e.g., synfuels or renewables) became available.

The third myth was that we were running out of oil. At projected levels of demand, reserves of oil would last less than forty years. This led to the concept of a developing gap between energy demand and supply. This myth was described by Adelman and Jacoby (1978) as the Mother Hubbard world view. One day, we would go to the cupboard, and Lo! It would be bare!

Closely associated with this myth was the fourth one, that somehow access to supply had in itself an intrinsic value. The view ran something like this: Oil is really worth more than its price. Therefore, I can purchase it only if I have a special relationship with the supplier. This view led to some extreme actions, some rather bizarre. Importers explored overseas, importers refused to trade with one another (thus strengthening the market position of the producer-exporters), oil companies bought coal companies, chemical companies bought oil companies, and oil companies bought one another. Vertical integration was now occurring *upstream* from the marketplace, whereas before it had occurred downstream from the production facility.

This myth led to the belief that a consumer needed to establish what came to be known as "assurance of supply," or "access to supply." The notion is that good relations with producers will ensure a purchaser a continuing supply when supplies are tight. This in effect assumes that suppliers will be willing to sell to friends at prices below market in tight market situations—even if it is against the supplier's best interest at the time. The point is, except for very short periods of time, no potential purchaser has been unable to obtain the supply desired, *if it was willing to pay the going price.* Hope of paying less than the going price in tight markets is simply commercial naiveté.

An additional myth was that the negative effects of energy conversion and use, such as those on the environment, were separate from the energy sector and could be easily addressed, particularly through technological solutions. The converse to this myth was that the energy crisis was so severe and so permanent that steps to end it would actually override environmental concerns.

The experts' scenario, of increasing energy prices and supply shortfalls, was based on long-term extrapolation of short-term phenomena. No one considered that oil was but one part of a system that would remain in balance, or that the way in which a balance might be achieved in the longer term was different from how it had to be achieved in the short term. In the short term the experts saw that the only practical response was to conserve energy by operational adjustment. Although directionally this was helpful, the strategy was—as we shall see—logically flawed in several ways. In the longer term, the experts saw a choice between filling the supply gap with alternative energies—whatever the cost—or permanent reduction in energy demand through recession or a deindustrialization of society, or both. This perception failed to recognize, among other things, the potential to increase the efficiency of energy utilization. Even more striking, it ignored the fact that efficiency would be the natural long-term response in a market economy, and therefore required no policy or special efforts or emphases. In the longer term it proved possible—even natural and evolutionary—to change operating practices and to alter or replace capital stocks and so achieve significant reductions in energy use without proportional reductions in output.

In short, the experts' paradigm was wrong and led to a misspecification of both policy objectives and appropriate policy actions.

## Policy Based upon Incorrect Economic Assessment

Based upon the experts' assessment, the emerging U.S. policy priorities for the post–1973 period were to fill the gap between demand and supply (largely with synfuels and additional use of coal), to attain energy independence by developing domestic energy resources, and to use newly developed domestic supplies to guarantee an eventual ceiling on the cost of energy, which was determined by technology. These priorities led to large-scale public and corporate investments in exotic energy technologies, for example, in the development of synthetic fuels, and to public investment in demonstration projects of renewable energies.

These investments were consistent with policies at the national level to achieve economic growth and energy independence. They were consistent at a corporate level with maintaining markets and revenues. But, for success, the investments depended critically upon the realization of the forecast increases in energy prices, without which the investments would be an unmitigated disaster.

### Policy out of Step with Societal Values

If our energy experts were ill equipped to analyze the situation post–1973, they were equally ill equipped to evaluate alternative policy options, either public or private. As well as getting the economics wrong, they wholly misunderstood the changing social climate in which their decisions would be judged. In the heat of the so-called crisis they overlooked that neither economics nor technology defines public acceptability. They merely define what is possible. It is society that dictates if what is possible is also acceptable, and society has the right to change its mind. Such a change can render prior technological or economic solutions irrelevant or impotent.

The 1960s, 1970s, and 1980s were decades of an emerging public consciousness and questioning: of the complexity of society, of a person's role in society, and of the appropriate role of technology. Public understanding of our environmental dependency and of the vulnerability of environmental systems took a quantum leap forward. We began to realize that there were upper limits to the amount of effluent and waste the environment could absorb without collapsing. We discovered that energy use and pollution are closely linked.

Since World War II, energy consumption had increased dramatically and total emissions of pollutants had grown apace. More alarming, the general shape of the energy consumption and emissions curves was exponential. By contrast, the absorptive capacity of the environment was stable or even shrinking. Equally troubling was the growing awareness of the health implications of atmospheric pollution surrounding conventional energy, and the waste and safety issues related to nuclear plants. Additionally, there was a growing feeling that we must avoid decisions that jeopardized not only the well-being of the current generation but also that of future generations. The notion that we were custodians of the earth led to demands for sustainable solutions. With this idea came a growing mistrust of high technology as a solution to our problems. There was a rising tide of skepticism about how successful previous technological solutions had actually been. There was a growing recognition that many problems were the side effects of solutions applied to earlier problems. Moreover, further technological solutions threatened to limit our options for the future by sending us into a spiraling dependence on technology.

For example, even if the envisioned synfuels solution had been successful technologically and economically, its massive capital demands would have resulted in even less energy flexibility. And the environmental issues it would have raised are enormous—the huge demand for water in arid states and the immense burden of disposing of the powdery shale waste, which weighs *ten times more* than the oil produced.

As one industrial executive observed (Byrom, 1978), hoping that more technology will fix the problems caused by technology is like hoping your spouse will solve the problems you wouldn't have if you weren't married.

The new consciousness implied that policy could no longer be based solely on economic considerations. The range of factors of relevance to society expanded from simple considerations of the cost and quantity of energy to include assurance and sustainability of supply, cleanliness, safety, and the long-term consequences of production and conversion technologies on the environment and human health. It was not that the energy experts overlooked the new dimensions, nor is it an indictment of them that they were unable to predict precisely how values would change. Many decisions that seemed optimal at the time, such as the commitment to nuclear power, were subsequently undermined by rapidly changing societal values. What the experts neglected to recognize was that major shifts in societal values were possible, likely, and crucial. Their oversight led to massive losses as projects that seemed both economically sound and technologically feasible foundered on the shoals of growing public resistance. The basic changes in the climate for decision making made it all the more important for experts to examine and reexamine their assumptions. Throughout the second phase, the real problem was the slow realization of how fundamentally different was the new energy reality. Experts failed to realize first, exactly how inadequate our analytical tools would prove and, second, how debilitating would be our lack of relevant experience. The outcome was that bad advice was followed, advice made credible because it came from an energy expert. It is not unusual for wrong advice to be given or followed.[7] The significant point is that errors made in the energy sector during the period prior to 1973 had limited repercussions. After 1973, the stakes were higher.

Although energy is a relatively small portion of the economy, it is absolutely crucial to the thing we call industrialization. Errors made in the application of conventional wisdom to other sectors did not carry either the high cost or the high visibility of errors made in the energy sector after 1973. What had been robust and transparent had become fragile and highly public. We made our biggest errors when we could least afford them—when new directions were being set for the entire industrialized world.

## THE THIRD PHASE: 1981–1989

What we are able now to learn from the first two phases runs something like this: in the short term, oil price ($P$) is inversely related to production levels, or supply ($S$). Before 1973, supply (i.e., production level) was determined by consumers (refiners-marketers of products). Thus, in the short term:

$$P = f(1/S) \qquad \text{Equation 2-1}$$

From 1945 to 1973,

$$S = f(\text{purchasers}) \qquad \text{Equation 2-2}$$

During the first phase, prices were stable because production levels ($S$) were constant and met demand, as determined by the international oil companies. Prices were low because production levels were high. Demand growth was dependable because price was constant.

The dramatic and irreversible structural discontinuity that occurred in late 1973 and after can be represented by:

$$S = f(\text{producers}) \qquad \text{Equation 2-3}$$

Since price remains an inverse function of supply, when the producers began to reduce production levels in late 1973, price went up. The supply reduction caused modern history's first oil price discontinuity.

For the next half decade, neither price nor demand changed much. Following the 1973–1974 price break, price didn't change much, because supply didn't change much. And, during that phase, *demand didn't change much, despite the higher price level.* That fooled us for a while, and contributed heavily to the gap mentality. What was missing was the realization that

People don't use energy—capital does.[8] Law 2-1

This principle was overlooked by those advocating conservation and a change in life styles as solutions. Certainly life style can affect marginally how much energy is used by capital, that is, by the hardware in place. Car-pooling certainly saves energy. But in many quarters measures were emphasized with inadequate appreciation of the time required to make significant changes in the capital stock in place, for example, to change the miles per gallon of the entire automobile fleet on the road.

In the short term, an increase in price doesn't change energy usage patterns very much. We may heat or cool our buildings a little less, or car-pool a little more, but the change in energy usage is small.

Big changes come mostly from replacing energy-using capital equipment and machinery with more energy-efficient equipment. New houses are built with more insulation; a Boeing 757 gets more air passenger miles per gallon of jet fuel; recuperators are added to boilers. But all that takes time. A nation's fleet of gas-guzzlers is not replaced with economy cars overnight just because gasoline prices increase. We wait until it's time to replace the old car; then, when we are ready, we buy one that gets higher mileage.

In the short term, price has only a modest, secondary effect on demand. Everyone knows that demand is inversely related to price; but, in the short term, the effect is small for energy. The significant impact of a change in price takes half a dozen years to show up.[9]

Therefore, the first corollary to our first law is that, in the long term—but only in the long term—demand ($D$) is inversely related to price ($P$). It is largely in the long term that:

$$D = f(1/P) \qquad \text{Equation 2-4}$$

It's our capital equipment that uses energy. Real energy savings require new capital equipment. And changing capital equipment takes time.

The second price discontinuity resulted from the decrease in production caused by the Iranian revolution and the Iran-Iraq war. The short-term dynamics—from 1979 to 1980—were just like those of the first price discontinuity. Prices jumped up. Soon thereafter, however, we noticed demand decreasing. This was unlike 1974. As our first law would have led us to expect, the demand decline beginning about 1980 was the result of the *first* price break (1973 to 1974), not of the second one! The new capital stock that the first price break had motivated us to order and install was just then coming into operation, and its effect on energy demand was becoming evident. *In a very real sense, it is the impact on demand caused by our investment in new, energy-efficient capital equipment and processes that marks the beginning of the third energy phase.* Therefore, the demand decrease of 1980 to 1985 was the result of the 1973 to 1974 price rise. (If *conservation* is redefined to consist of this economically driven investment in new, energy-efficient capital equipment and processes, then one could say that the fruits of "conservation" mark the beginning of the third energy phase.)

Appearances masked the realities. The appearance was that demand increased in response to the 1973–1974 price rise, and decreased in response to the 1978–1979 price spike. In reality, the demand response was lagging price behavior by six years. The reality was that the demand increase following the 1973–1974 price spike occurred despite the price rise. The demand decrease following the 1978–1979 price rise was the result of the 1973–1974 price rise. But at the time the phenomenon was not widely understood.

The reduction in OPEC production in 1973–1975 caused a price increase beginning in 1974 and continuing on a plateau through 1978. (See Figure 2-4.) Despite the higher price, free world demand grew through 1979. (The temporary decline in free world demand in 1974 and 1975 is more attributable to the worldwide recession than to oil prices.)

The sharp decline in OPEC production in 1979–1980 caused the second price increase in 1979, which continued through 1982. The higher price could not be sustained, since free world demand fell consistently starting in 1979 *because of the higher prices starting in 1973.* OPEC production had fallen also during this phase, but not fast enough—compared to the falling demand—to maintain prices or prevent their decline.

An additional factor, the magnitude of which is easily overlooked, is the long-term effect of the higher price levels on non-OPEC supply. Like the response of demand to price, the supply effect also has a significant time-delay element. As prices rose from 1973 to 1976 non-OPEC free world oil production actually *fell.* (See Figure 2-5.) It has risen steadily since then, as a result of the higher price levels since 1973. By 1985, it had increased seven million barrels per day,

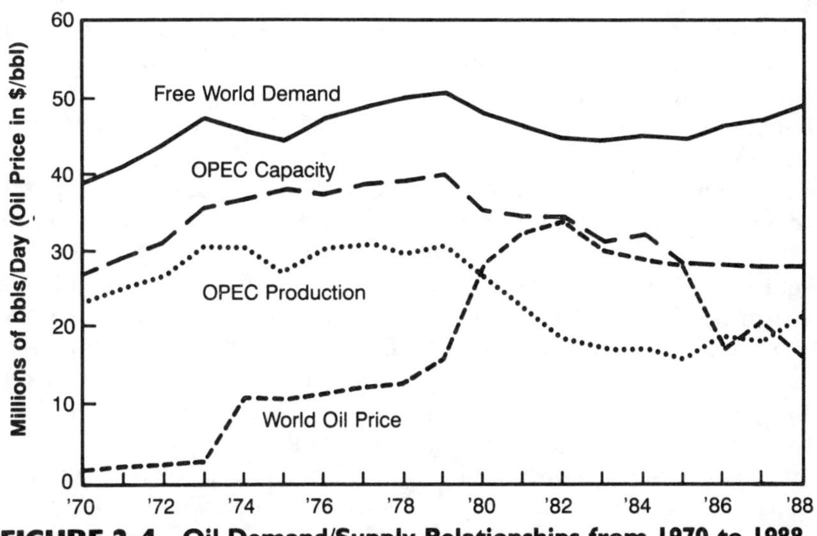


**FIGURE 2-4 Oil Demand/Supply Relationships from 1970 to 1988.**

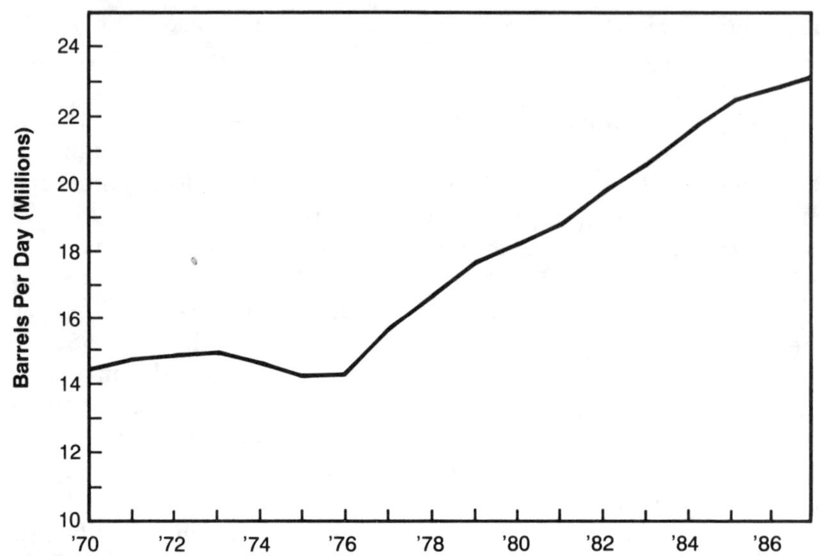


**FIGURE 2-5 Non-OPEC Free World Oil Production from 1970 to 1987.**

or about one-fourth of OPEC's peak production. Not only does the absolute volume bring immense downward pressure on price; but even more important the *new* supply fragments the world oil market, and introduces new forces with diverse interests and perspectives. It also introduces new actors, who have not yet become established and accepted members of the producers' "club." The resulting lack of coordination of production levels with ultimate product demand has contributed greatly to what is new (since 1973) to the world oil market: intrinsic instability.

Finally, the data emphasize the inappropriateness of the popular premise that we are running out of oil. Except for 1979 and 1985, the unused OPEC capacity has *grown* steadily from 1970 to 1986. (See Figure 2-4.) This is a valid measure, at least midterm, of whether we are running out of oil, and of the value—if any—of an ensured supply. If the steady growth in unused OPEC production capacity kills this myth, other data bury it: non-OPEC production increased 50% from 1976 to 1986. (See Figure 2-5.) The delayed demand reduction, then, made the 1979 price break extremely unstable.

The key to maintaining a crude oil price level is to control production levels so that they just meet demand. To do this means gaining a concomitant control of surplus production capacity, so that it does not erode prices. Prior to 1973, control was accomplished in the United States through the Texas Railroad Commission; in the world oil market, it was accomplished through the company control of production levels in exporting countries. From 1973 through 1980, it was accomplished through discipline within OPEC. To sustain the relatively high price levels of 1979, the exporters, who then controlled production levels, would have had to reduce supply by an amount greater than the significant and unexpected reduction in demand, plus the additions to supply reflected in Figure 2-5. This they did not do, and prices eroded beginning in 1982.

In the short term, price is inversely related to supply. In the face of decreasing demand, supply didn't decrease. So, price did. (See Figure 2-4.)

Then came 1986. Half a dozen years after the Iranian revolution, we witnessed another reduction in demand. The reduction was a response to the second price spike from 1979 to 1982. To sustain the prices of the early 1980s, the exporters would have had to reduce supply by the same amount. This they did not do, resulting in the third price discontinuity: the first price plunge in half a century!

## DISCONTINUITIES AND DECISION MAKERS

Price behavior during the last fifteen years has been the antipode of trends, and confounded our ability to forecast. The structural discontinuity laid the groundwork for the three price discontinuities that followed. The discontinuities have been separated by periods of six years or so. In between the discontinuities, price has been relatively stable. Understanding price behavior is a matter of understanding the *discontinuities*, not of discerning or projecting *trends*.

We can understand the dynamics, and we can predict that there will be more discontinuities in the future. But forecasting price is clearly an exercise in futility. Far worse, it misleads us. The fact is, energy prices in the midterm are uncertain over such a wide range that forecasting is simply not useful.

The inability to forecast price leaves the decision maker in a real quandary. For example, let's assume that our first law continues to hold. The lower oil prices we have seen these last few years are causing capital replacements to be less energy efficient than they would have been if oil prices had been higher. The result will be renewed growth in demand beginning six years after the start of the slide in prices that began about 1981. That should make it about 1987. And that is just what happened. It is largely in the long term that:

$$D = f(1/P) \qquad \text{Equation 2-4}$$

In the winter of 1988–1989, that is exactly what happened. In addition, the concomitant occurrence in supply could also be observed: during the winter, U.S. petroleum imports exceeded U.S. production.

These phenomena clearly illustrate the inevitable whipsaw effect resulting from the long delay in the response to changes in oil prices, caused by the capital intensity on the demand side. Of course, the whipsaw is exacerbated by the effect price has on exploratory and developmental efforts worldwide. This effort also has a delay of six years or so. That is, the exploratory effort spurred by the high prices of the late seventies is just now coming on stream, when the world doesn't need additional production capacity. Since most of the new production is outside OPEC, the OPEC countries are forced to take the reduction in production. OPEC's share of the supply of free world demand fell from two-thirds in 1973 to one-third in 1985. (See Figure 2-4.) The increase in non-OPEC production made up the difference. (See Figure 2-5.)

Unfortunately, it is not as simple as a "six-year cycle." For the demand response depends not only on unpredictable price changes but also on changing industrial standards, the varying impacts of memory on capital decisions, the differing life of capital stock in place, and so forth. Energy demand at any time is a result of the cumulative effects of all forces. As the price instability continues, the cumulative effects become increasingly difficult to analyze and quantify. The delay between price change and demand change is a big contributor to the difficulty of analysis.

The system itself perpetuates price instability. Indeed, if the exporters succeed in their efforts to constrain supply, the next few years will be interesting. World oil prices will continue to be unpredictable and uncertain for the foreseeable future, that is, for the rest of the century. On the supply side, the increase in the diversity of production, such as new sources in the North Sea and offshore Yucatan, tends to attenuate the volatility in an upward direction but to exacerbate it in a downward direction. That is, when prices are rising, new sources will dampen the rise. But when prices begin to fall, the same new sources will accelerate their decline.

On the demand side, the natural economic response to the two price spikes of the 1970s (e.g., increased flexibility in fuel choices and increased energy efficiency in capital stock) adds needed stability to the world oil market.

## The Range of Prices

Prices now move, not along trends, but in discontinuous spikes, breaks, and plunges. How wide is the likely range of these discontinuities?

Already, since 1974, we have seen prices range from $12 to $48 per barrel, in current dollars. A factor of four! In 1988 dollars, the range is nearer a factor of five! So, we know from fifteen years of history that the range can be at least that great. Even that range bumped neither the lower nor the upper limits of real possibility. The lower limit is around $1 to $4 per barrel. Anyone can argue about the exact figures, but most will agree that in the medium term—say, several years—there is some thirty million barrels per day of production capacity available to the world market at a marginal cost of $1 to $4 per barrel (1988 dollars). We mean marginal in the midterm; that is, these figures include operating production costs, but not finding costs. This oil is largely from the Persian Gulf, Africa, the South China Sea, Mexico, the North Sea, and South America. The volume is adequate to supply the world oil market in the medium term, and thus determines the lower limit of world oil price in the medium term. (This discussion assumes that higher-cost indigenous production would not be reduced significantly in the midterm. In practice, a number of mechanisms are available to guarantee such a result.)

The upper limit is at least $65 per barrel in 1988 dollars. This price corresponds to the $50 per barrel in (1982 dollars) that purchasers were in fact willing to pay in 1982. (Calculations indicate that the consuming world would be willing to pay almost twice that figure rather than do without imported oil.) Consumer attitude determines the upper limit of world oil prices in the medium term. Therefore, for the next fifteen years, the range of real possibilities for world oil price in the medium term (i.e., a few years) is conservatively estimated as a factor of thirteen (say, from less than $5 to more than $65 per barrel). The intrinsic instability and unpredictability are characteristic of the third energy phase. In retrospect, we now know that this was also true of the second phase.

Our prediction of "instability and unpredictability" may seem as foolish as trying to predict prices or stability. We do not predict, however, that future discontinuities are inevitable. We do argue that attempting to predict either specific changes or specific trends is futile. Prices will either rise, fall, or remain level. Our argument is that it is beyond the capacity of analytical powers to predict which level, within a useful degree of accuracy.

For example, government forecasts from 1979 through 1981 have been dramatically wrong. (See Figure 2-6.) Not only were the price forecasts made by the prestigious U.S. Department of Energy's Energy Information Agency ludicrous; but also, the forecast for 1985 made in 1977 was more accurate than that made in 1980! We don't even get smarter about the future as we get closer to it!

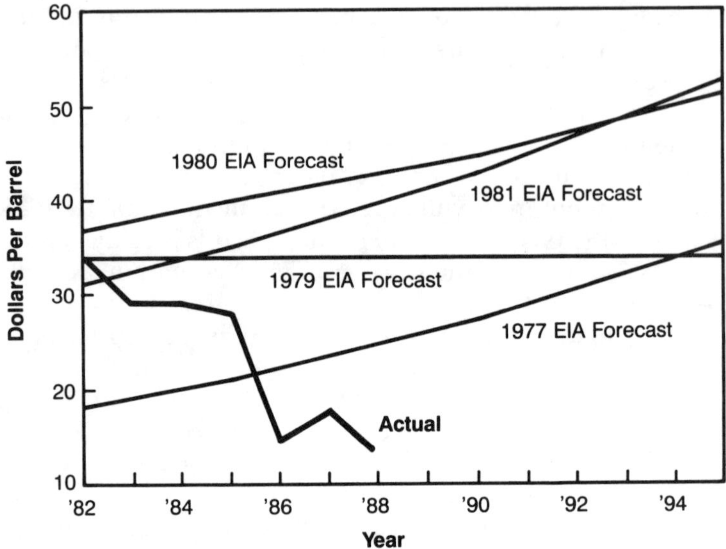


*Source:* Energy Information Agency, Department of Energy.

**FIGURE 2-6 EIA Price Forecasts for 1977, 1979, 1980, and 1981 (constant dollars).**

What marks the third phase is that now we *know* that the long-term instability and unpredictability is systemic. The real danger is that we may mistake a few years of relative freedom from trauma for equilibrium. What we can learn is that, at best, what we have is a few years in which to reflect on our blunders, learn from them, and prepare for the future.

## CONCLUSIONS

At the time when we most needed clear, well-founded, and constructive advice, our energy experts were most lacking, and we have considered some of the reasons: the dramatic switch in the energy situation and the changing values and expectations of society. By themselves, these reasons do not explain why mistakes were made. Certainly, the climate for decision making had changed radically, but the real explanation for our mistakes was that there was so much wrong with our analytical and decision-making methodologies, and perhaps more significantly, we placed too much faith in them.

In 1973, based on our experience during the first phase, we began our analysis by addressing the wrong issues. The questions we asked conditioned our way of thinking and constrained our imagination about the potential solutions. We sought energy independence, ways to fill the energy gap, synthetic replacements for oil and natural gas, and access to supplies. Our forecasts were invalid. We virtually overlooked the basic economic principles that would still operate to regulate energy prices and balance supply and demand. The models we used to test policy options were based on false assumptions and inadequate measurements. They were oversimplistic, and yet we had faith in the output. Our cost-benefit analyses were too narrowly prescribed, recommending solutions that were suboptimal from a societal viewpoint or longer-term perspective. Above all, there was a belief in the power of technology to yield solutions and a faith in the capacity of policy measures to influence private decisions. Both were faulty.

We devote the remainder of this book to a detailed analysis of the mistakes that were made at each step in the decision-making process. We draw on a substantial body of research and case histories of important decisions to illustrate our conclusion that many of the methods used to analyze energy issues, and most of our cherished beliefs and views about energy, were either inappropriate or simply untrue. In so doing, we not only explain how so many mistakes were made but also draw out the lessons, which point toward a more accurate assessment of energy in our future and a more realistic path for development.

## NOTES

1. Exxon, Mobil, Texaco, Gulf, Chevron, Shell, and British Petroleum. All these companies are U.S., British, or Dutch. The picturesque description, *le sette sorelle*, was first applied to them in the 1950s by Enrico Mattei. Mattei was a pioneer in meeting these firms on their own ground—control of third world production of crude oil—through his Italian firms, Ente Nazionale Idrocarburi (ENI) and Azienda Generale Italiana Petrole (AGIP).

2. We use this term in its technical economic sense: the difference between the *price* of oil and its *value* to the purchaser. As we were to learn in the fall of 1973, what the oil companies had been selling for $2 per barrel, the purchasers were willing to buy at $12 per barrel—and in 1979, $35 per barrel. The difference between the $2 and the $12 or $35 is a portion of what is meant by "consumer surplus." Of course, the value to the purchaser (in these examples $12 and $35) is a function of both the quantity and the kind of sunk investment in oil-using capital equipment. Therefore, the concept of consumer surplus is useful only in the midterm, that is, in the time it takes to turn over substantially the oil-using (and oil-conserving) capital equipment. As we shall see, this turnover takes five to ten years.

3. There were, in fact, some warnings, for example, in the U.S. Senate's "National Fuels and Energy Policy Study," 1974. What was missing was public credibility of an approaching problem.

4. The Organization of Petroleum Exporting Countries is now composed of Algeria, Ecuador, Gabon, Indonesia, Iran, Iraq, Kuwait, Libya, Nigeria, Qatar, Saudi Arabia, the United Arab Emirates, and Venezuela.

5. As we shall see shortly, the next shift, in 1981, was also qualitative. At that time OPEC lost control of production allocations. Although each exporter retains control over its own production, there is no effective control over the total as there was in 1974–1980. So, it is fair to say that today no one is in "control." We have the oversupply and soft prices one would expect for a mature commodity with uncontrolled excess production capacity.

6. One of the authors has long argued (see, for instance, Ball, 1985) that the exporting countries now have the same reasons to integrate downstream that the integrated companies had many decades ago. Similarly, the reasons the integrated companies originally had for integrating no longer exist in most parts of the world. Therefore, a return to stability in the world oil market might lie in the direction of a metamorphosis in vertical integration, from the companies to the exporting countries. In fact, some tentative steps are being taken in the form of the sale of downstream assets located in importing countries by integrated companies to exporting countries. Examples include the sale of refineries, distribution systems, and retail outlets in the United States and continental Europe to exporting countries.

7. Robert A. Millikan (winner of the 1923 Nobel Prize for physics) and, at one stage, Albert Einstein both believed that it would never be possible to tap the power of the atom. In 1932, Einstein argued that "there is not the slightest indication that [nuclear] energy will ever be obtainable. It would mean that the atom would have to be shattered at will." Einstein subsequently revised his opinion about the probability of harnessing nuclear energy and by 1955, Gen. David Sarnoff, chairman of the board of the Radio Corporation of America, was forecasting that atomic batteries would soon be commonplace. "It can be taken for granted that before 1980 ships, aircraft, locomotives and even automobiles will be atomically fuelled." This optimism was shared by both Henry Luce, founder and publisher of *Time*, *Life*, and *Fortune* magazines, and John von Neumann, Fermi Award–winning American scientist, who, in 1956, independently prophesied a future with costless or near costless energy. Cerf and Navasky, 1984, pp. 210–211 and 214–216.

8. For a more extensive description of the principle, see Ball, 1986.

9. We are indebted to our MIT colleague Prof. M. A. Adelman for the concept.

## REFERENCES

Adelman, M. A., and H. D. Jacoby. 1978. "Oil Gaps, Prices and Economic Growth." MIT World Oil Project. Working paper number (May) MIT-EL-78-008WP. Cambridge, MA.

Ball, Ben C., Jr. 1985. "Management Strategy for the Takeover Decision." *Petroleum Management* (August), pp. 27–32.

———. 1986. "Oil Prices and the Future of the Petroleum Industry." *Petroleum Management* (November), pp. 17–56.

Blair, John M. 1978. *The Control of Oil.* New York: Vintage Books.
Byrom, Fletcher. 1978. Chairman of the Board of Koppers Corporation. Pittsburgh, PA, personal communication.
Cerf, Christopher, and Victor Navasky. 1984. *The Experts Speak.* New York: Pantheon.
Sampson, Anthony. 1975. *The Seven Sisters.* New York: Viking.
United States Senate. 1974. "National Fuels and Energy Policy Study." S442-21.

# 3 Government Blunders

This chapter describes the U.S. government's assessment of the changing energy situation, its relatively passive role during the first energy period, and its popularly supported decision to intervene with a heavy hand at the start of the second energy period. By way of illustration, we detail governmental policy on pricing, conservation, synfuels, photovoltaics, renewables, natural gas, and nuclear—outlining the blunders made in each area of policy during the second energy period. We also describe the gradual withdrawal of government from active participation in the energy sector during the third energy period.[1]

## THE SCHIZOPHRENIA OF OIL PRICING

From 1945 to 1973, the U.S. government played a relatively passive role in the oil sector. The United States differs from most other nations in an important aspect: here mineral rights are attached to surface rights. In most of the rest of the world, all natural resources belong to the national government.[2] Although it is true that an appreciable portion of the U.S. energy resources lies under federal lands (or waters), the U.S. government treats the property much as private landlords do—that is, it leases its mineral rights to the high bidder. In no case is it the operator. It receives a fee for the lease and a royalty for the oil. It acts as neither a producer nor a seller of oil. Leasing takes the U.S. government out of the energy business as a direct participant.

Because the government is neither a producer nor a consumer of energy in any significant amount, it can only affect the energy market indirectly, through broad policy tools. It can only provide the context and sanctions within which market decisions are made. It can set ceiling prices on gasoline, but it cannot *decide* how much gasoline will be used. It can forbid new utility boilers to burn natural gas, but it cannot *decide* how much gas will be used. It can affect the market dramatically, but it cannot determine the market. For example, it can fund solar research. It can even offer tax incentives for the use of solar devices. But it cannot create, require, or supply a significant market for solar energy devices.

Before 1973, the world petroleum industry was dominated by U.S. firms and U.S. technology. U.S. oil taxes were low, compared to the rest of the world, so U.S. consumers had relatively low energy prices. This was true even though U.S. crude oil *costs* were higher than those in the other industrialized countries.

Most of the oil consumed in the United States was produced domestically, at a price of $3 per barrel. World oil prices were about half that. This margin could be maintained because of federal quotas on lower-priced imported oil.

Part of the reason for the discrepancy is political and part is geological. One political reason is the private mineral rights just discussed. Since the mineral rights match surface rights, the only way for an individual to get "his" oil was to have "his" well. (Prior to U.S. offshore production, which became significant only relatively recently, most U.S. production was on private, not federal, lands.) Thus wells were drilled in the context of surface ownership, rather than cost effectiveness. To a large extent, each owner had to drill a well in order to get oil from under his land. The result is that many more wells were drilled than were needed technologically.[3]

A second political reason is that the quintessential oil regulator, the Texas Railroad Commission, whose pattern was followed in the large producer-states, set "allowables" on the basis of so many

**TABLE 3-1 Rates of Production by Country and Well**

| Country | Crude Oil Production (bbl. × 1,000,000) | Producing Oil Wells | Daily Average Production per Producing Oil Well (bbl.) |
|---|---|---|---|
| United States | 3,168 | 635,015 | 14 |
| Canada | 537 | 36,955 | 40 |
| Venezuela | 653 | 9,971 | 180 |
| Malaysia | 243 | 923 | 720 |
| Mexico | 913 | 3,263 | 766 |
| Nigeria | 537 | 1,127 | 1,304 |
| Saudi Arabia | 1,801 | 592 | 8,336 |
| Iran | 704 | 221 | 8,728 |

*Source:* Adapted from DeGolyer and MacNaughton. 1988. *Twentieth Century Petroleum Statistics*, Dallas, TX, p. 1.

barrels per day *per well.* Therefore, there was a great incentive to drill more wells. The saying was, "You drill wells not to get oil, but to get allowables."

The additional reason for the higher cost is simple geology. Compared to other major oil provinces, U.S. fields intrinsically yield a low production rate per well. All of these factors combined have dramatic consequences. (See Table 3-1.)

The daily average production per producing oil well is inversely proportional to the cost of the oil produced.[4] The United States—along with Canada—thus stands in a unique position among industrialized nations of the free world. Before the 1973 price spike, and before North Sea production became significant, these two countries were the only industrialized ones in the free world with significant production relative to demand. Their indigenous production, however, had a higher cost than the oil imported by the other industrialized countries. Nevertheless, public policies resulted in lower consumer prices for oil products in the United States than in the other industrialized countries.

In the United States, the lower price was solely because of the absence of comparable excise taxes and duties, and the existence of tax breaks for producers. We refer, of course, to the famous depletion allowance and foreign tax credits. Regardless of any merit either might have had from a viewpoint of public policy, the final consequence was lower prices, and not increased after-tax profitability. The evidence for this is the mediocre return of the major international oil companies, compared to those of the mining and manufacturing sectors.

As imports into the United States rose during the 1950s, their volume came under control for the expressed reason of national security. The economic value equal to the difference between the world price and the domestic price went to the importer. The price to the domestic buyer was set by domestic price, not by the price of imported oil. High-cost domestic production was protected against low-cost foreign production by quotas, but low import and excise taxes kept prices to U.S. consumers low relative to those for European consumers—even though European petroleum products were refined from the lower-priced imported oil. Thus, even prior to 1973, the United States lived in an energy world different from that of most of the rest of the world. Although the U.S. government had myriad policies that impacted the energy sector, it had no explicit and cohesive energy policy, in the sense that one could say we had a defense policy, where purchases and sales are in fact made by the government, and the government is a significant player in the marketplace. A *qualitative* difference should be noted here. The government has direct, executive, managerial, and budgetary control over military issues. By contrast it neither buys nor sells significant amounts of energy.

The oil industry, both in the United States and worldwide, was dominated by U.S. firms. We were rich not only in indigenous oil production but also in gas. Unlike any other country in the world (except Canada and Russia), our economy was affected by the economics of oil production as well as the economics of oil consumption.

Prior to 1973, government involvement in the energy sector was focused on the two tax provisions already mentioned (depletion allowance and foreign tax credits), import quotas (beginning in 1958), price controls (on gas beginning in 1954, crude and products beginning in 1971),[5] and, of course, electric utility rates (federal and state regulation) and the nuclear fuel cycle. In addition, there was the regulation of interstate pipelines and the leasing of government-controlled mineral-bearing properties. Other government policies had an effect on the energy world:

- favorable tax treatment at the federal, state, and local levels for public power; e.g., state power authorities, municipally owned electric utilities; rural co-ops; and TVA;
- subsidized construction of major hydropower projects by the U.S. Army Corps of Engineers and the Bureau of Reclamation; and subsidized power to public power companies;
- subsidies in the form of low-interest loans, loan guarantees, and financing through federal guaranteed loans;
- subsidization of the federally owned and operated uranium enrichment complex;
- a limit to the liability of electric utility companies in nuclear accidents through the Price Anderson Act of 1957;
- numerous other federal subsidies; e.g., forgiveness for the coal industry of $3 billion in interest on funds borrowed from the Treasury by the Black Lung Trust Fund; forgiveness of penalties on prepayment of loans by the rural electric co-ops, thus permitting refinancing at lower interest rates and costing the U.S. Treasury billions of dollars.

The government's indirect involvement had some powerful and subtle influences. Examples would certainly include those that promoted the automobile and airplane over the train and other forms of public transportation. Among them were:

- low-cost and easily available FHA and GI mortgages, and tax deductions for mortgage interest (as opposed to nondeductible rent), which promoted suburbia and the concomitant commuter traffic;

- low taxes on automobiles and tires, which enabled multiple-car families to become the norm;
- the building of the interstate highway system, which made automotive travel pleasant and truck transportation the rule; and
- the transfer of mail subsidies from rail to air, which helped hasten the decline of much of the U.S. railroad industry.

The result, of course, was an unconscious but intense, long-term and essentially irreplaceable reliance by the United States on plentiful, low-cost gasoline—a reliance far greater than that of any other nation. For example, half of our crude is refined into gasoline—about twice the percentage of most industrialized nations. (See Table 3-2.) When our gasoline supply seems threatened, or when its price escalates, we are in real trouble—far more so than any other country. The dependence on gasoline is exacerbated by the fact that both technologically and economically, gasoline is the most difficult fuel for which to find substitutes on the supply side and the most difficult fuel to substitute for on the demand side. That is, there are no economically and technologically adequate substitutes with the qualities of either gasoline or the gasoline engine. All attempts at substitute fuels have failed, mostly on economic grounds. Attempts have included natural gas, propane, butane, methanol, ethanol, alcohol-gasoline mixtures, and even hydrogen and electricity. Any limited apparent commercial success by these fuels has been possible only because of special tax treatment or subsidies, or both. For example, the use of liquefied petroleum gas (LPG—liquefied propane or butane) as a motor fuel in Japan illustrates that special financial incentives are needed to make substitute fuels viable.

By contrast, natural gas is—where available—a reasonable

**TABLE 3-2 Largest Users of Gasoline in the Free World**

| *Country* | *Demand for Gasoline as Percentage of Total Refined Products* |
|---|---|
| United States | 43 |
| United Kingdom | 29 |
| West Germany | 24 |
| France | 23 |
| Italy | 17 |
| Japan | 15 |

*Source:* Adapted from DeGolyer and MacNaughton. 1988. *Twentieth Century Petroleum Statistics,* Dallas, TX, p. 15.

substitute for residual fuel in an industrial or utility boiler; and, in the medium term, an industrial or utility user can easily switch from oil or coal to gas.

In the United States both crude and derivative products fell under general price controls in 1971, but during the first two years the effect was only slight. The U.S. price was insulated from the world price by the import quotas. When, in late 1973, world prices skyrocketed past the U.S. price, the quotas became an anachronism. But price controls on crude and products were continued past the end of general price controls. With world price climbing to ten times and more the base U.S.-controlled price, the effect of price controls was no longer slight, but very dramatic. At the very time we should have been afforded the economic incentive to reduce oil consumption, we were "protected" by our own government from experiencing the economic incentive. The U.S. consumer and the U.S. economy did not feel the full effect of the spiraling world oil prices until decontrol of crude oil prices in January 1981! Thus, unlike the rest of the industrialized world, the United States had low oil prices before 1973 (despite high costs) because of U.S. tax and import policy, and from 1974 to 1981 because of price control.

During that latter period, of course, the marginal cost of oil was the world price. But the workings of the federal entitlements program ensured that consumers never saw the marginal price but were aware only of the average price. The intent was to equalize the cost of oil to all refiners, whether they purchased U.S. price-controlled oil or higher-priced imported oil. The equalization in costs among refiners, of course, equalized the price they charged for products. Whereas before 1973 the import quota program insulated U.S. consumers from the relatively lower world price, after 1973 the entitlements program insulated them from the relatively higher world price. In 1981, for the first time in a quarter of a century, U.S. crude price had to compete with world crude price without huge artificial barriers. U.S. crude prices equaled world oil prices. For thirty-five years, oil prices paid by U.S. consumers had been significantly lower than those paid by consumers in other industrialized countries. This lower price continues even today, because of the consistently lower U.S. excise and import taxes. The result is, of course, higher energy use in the United States than in other industrialized countries.

When price controls on crude oil ended in 1981, they were replaced with a tax erroneously called a windfall profits tax. This is not a tax on profits, but an excise tax. The intention of the tax was to capture for the government (the people) the revenues attributed to the run-up in world oil price. Moral virtue aside, its effect has been to constrain to some degree exploration, development, and production of crude oil in the United States. Incidentally, whatever value a depletion allowance might have had in stimulating indigenous production was nullified in the mid-1970s when the program was—for practical purposes—terminated. This was just at the time national policy was calling for increased indigenous exploration and production.

Thus, stated national policy has been to increase indigenous supply; but from 1973 to the present, de facto policy has, in a number of important ways, inhibited it. Stated national policy has been to restrict energy usage; but until 1981, de facto policy has encouraged it.

## CONSERVATION

The oil price shock of 1973 created in its wake a potpourri of slogans, public awareness programs, and other efforts to reduce U.S. dependence on imported oil, especially OPEC oil. As a nation, the United States developed intensive programs to reduce consumption of oil and other forms of energy in all aspects of life, from automobiles to industrial processes to household usages. The programs produced news articles, bumper stickers, and emotional heat, but they were dysfunctional to an understanding of the energy problem.

An error common to the programs was the concept that it was wrong to consume, rather than that we should consume *wisely* in view of the higher price of energy. For example, a goal was that we should consume less, even where less meant also less comfort, less productivity, and fewer goods and services—*regardless of the cost effectiveness.* The mistake was in presuming that consuming less *energy* was itself the goal, and that the goal had an intrinsic value. The blunder lives on today in the mandates of virtually all state energy agencies.

We now realize that energy is always available at a cost. What we are interested in is not in reducing the amount of *energy* that we use, but rather in minimizing the total *cost* of our enterprise. Leaving the question of social cost to a later point in the book, that total cost can be represented by the classic equation:

$$C = K + L + E + M.$$

The total cost $C$ is the sum of the costs of capital ($K$), labor ($L$), energy ($E$), and materials ($M$). This is as true of building a ship or heating a building as it is of driving to the store in the family car. Being economic creatures, we naturally attempt to minimize $C$, and so arrive at a certain "recipe" or mix of $K$, $L$, $E$, and $M$ that minimizes $C$ for any activity. However subjective their measurement must be, convenience and comfort are most certainly included in our evaluations, for example, of $K$ and $L$. When the cost of $E$ jumps substantially, the recipe needs to be readjusted. The mix needs to be reoptimized, replacing some $E$ with capital, labor, or materials. Thus, we insulate, we drive more slowly to obtain better gasoline mileage, or we return to using a solar clothes drier (a clothes line). We make choices that minimize total costs. It would be as foolish to minimize energy or even reduce it by an arbitrary amount as it would be to minimize, say, capital, or reduce it by an arbitrary amount.

Our failure to understand this important principle is illustrated by President Carter and his advisers' harangues about how *wasteful* Americans are compared to our European counterparts. He quoted numbers for BTUs used to make a loaf of bread and a ton of steel in the United States and in various European countries. Of course, the United States used more energy. But with the government keeping our energy prices lower than Europe's as a matter of public policy, it isn't difficult to figure out why. With cheaper prices, our "recipe" is different, employing more energy and less capital, labor, and material than Europe's. To have done differently would not only have been foolish; it would have meant noncompetitiveness—economic disaster. A more appropriate message would have been: "Prices have changed, so be sure to reformulate the recipe accordingly." We made conservation a *moral* issue when the real issue involved an appropriate allocation of economic resources. Not only was the energy crisis not the "moral equivalent of war," it was not the moral equivalent of anything. When policymakers confuse moral issues with economic issues, they make bad policy.

It is poor public policy, for example, to give a tax credit on energy-saving investments if the only purpose is to encourage conservation. If the investment is sound economically, then it needs no tax incentive. If the investment is sound only because of the tax incentive, then the tax incentive represents simply a transfer payment motivating an uneconomic expenditure, and is clearly bad public policy. Admittedly, the argument can be made that such a tax credit could be *good* public policy if it offset other market dislocations or distortions that obstruct efficient decisions. In practice, however, this often smacks of "two wrongs make a right."

In addition, if something is very cheap, then wasting it is not so bad; saving it may literally not be worth the trouble or expense. For example, a certain amount of car-pooling makes purely economic sense at almost any price of gasoline. But for many people, car-pooling is an inconvenience. Therefore, the savings in gasoline must be weighed against the inconveniences of car-pooling. As gasoline prices increase, more individuals will naturally opt for car-pooling. In effect, there is a gasoline price at which the inconvenience becomes worth the monetary savings.

A different kind of example may serve to illustrate the important principles even more clearly. If water is cheap, it may be more expensive to repair the plumbing than to endure the drip. However, if the price of water escalates dramatically, then replacing the plumbing will, at some point, make good economic sense.

Two important points can be emphasized using a commonplace example. The first is that there are supraeconomic, subjective, psychological, nonquantifiable factors that can play a large role in decision making. In the case of a leaky faucet, part of the decision to fix it or not involves subjectively making the nonquantifiable trade-off between the annoyance of the sound of the drip on the one hand and the bother of hiring the plumber on the other hand. How this trade-off is actually made will vary dramatically from one individual to another, depending on the values for each of the supraeconomic factors involved. The actual price paid for the water and the plumbing repair are only the most obvious and easily quantifiable factors. To any one individual, either factor may not even be the most important.

The second point to be made is that the price charged to the customer may well not represent the total, long-term cost to the economy or the society as a whole. For example, it may not include the societal cost from the water table being lowered, or the cost of the environmental damage caused by runoff or sewage. Nevertheless, the customer is led to make decisions—in effect, to conduct a cost-benefit analysis—on the basis of the price charged, not the total, long-run, social cost.

A buyer responding to the price of energy could well be better off using less energy by switching some energy expenditures to capital expenditures—a change in the inputs to the final demand rather than the *quantity* of final demand. An industrial firm faced with the higher oil prices of 1975 was fully cognizant of the desire for energy conservation but was more motivated by good business than by governmental jawboning to consume less energy. As long as the two objectives netted the same behavioral change, either side can claim victory. Had using less energy (i.e., reallocating resources from energy) not been in the best economic interest of the firm, it would probably have continued its past practices. Certainly in energy-intensive businesses, reallocating resources toward capital and away from energy was essential to maintaining a competitive position. Two brief examples highlight these points.

## Conservation Tax Credits

The original Energy Policy and Conservation Act was signed into law by President Ford on December 22, 1975.[6] This act and its follow-ons, the Energy Tax Act of 1978 (PL 95-618) and the National Energy Conservation Policy Act (PL 95-619), which were part of the Carter energy package, provided the legal structure for tax credits for conservation-related investments. PL 95-619 provided for funding of programs for

- Title II: Residential energy conservation
- Title III: Energy conservation programs for schools and hospitals and buildings owned by units of local governments and public care institutions
- Title IV: Energy efficiency of certain products and processes
- Title V: Federal energy initiatives
- Title VI: Additional energy-related measures
  - Industrial energy efficiency reporting
  - State energy conservation plans
  - Minority economic impact
  - Studies
- Title VII: Energy conservation for commercial buildings and multifamily dwellings

Programs were mandated to get conservation off the ground in both private and public buildings. Market forces were *declared* to be too slow, thus requiring the government to step in. Combined with PL 95-620, PL 95-619 became the carrot motivating consumers to move. Qualifying energy conservation expenditures in insulation were subject to rebates of 15% up to $2,000. Renewable energy systems received rebates of 30% up to $2,000 and then 20% for the difference to $10,000. The incentives were there. Now it was up to the marketplace and public information to make them happen.

But would the majority of these private-sector investments have happened anyway? Who benefits from programs that provide tax subsidies for conservation investments? In 1978, the clear popular answer was that society benefits; so society, through the government, should make them. The counterargument in the case of energy conservation expenditures is that the proper ones would have been made anyway based simply on the signals of the market, and on the economic self-interest of consumers. One should not forget that, during much of this time, price controls on both oil and gas gave false market signals. Prices were artificially depressed, thus encouraging the use and discouraging the "conservation" of these energy sources.

No one likes paying outrageously high utility bills. Even the residential homeowner can see that simple caulking, insulation, and storm windows with three-year paybacks are worthwhile investments. The rational homeowner will make the decision regardless of the tax credit. The tax credit just shifted the burden of payment from the beneficiary to society! Admittedly, tax credits *did* serve the function of calling attention to an issue that had heretofore not been important.

## Mass Transit

The mass transit system is a positive force in energy conservation.[7] This statement seems sensible. The corollary seems equally sensible. Extend the system and you will save even more energy. How? Pick up the commuters in the suburbs and bring them into the city. Although on the surface these statements appear logical, they are, under a wide set of conditions, incorrect. Here are three statements commonly accepted as true, and a closer look at why they are false.

- Mass transit is always energy efficient. The majority of mass transit systems are based on buses. An average bus must have a loading factor of greater than 25% before it is superior, from an energy perspective, to driving a small private automobile with two occupants. In most urban areas these values are not met except on high-density routes.
- Extending mass transit always conserves energy. Generally mass transit systems require more energy per mile as the extensions move into less-populated areas. Fewer people, not more, ride; *and* the bus is less likely to go where people want to go.
- Extension of the mass transit system into the suburban areas is economic because people want to go from the suburbs to the city to work. This statement is not supportable in the United States, and is becoming shaky even in other developed countries. People are living wherever they wish. The majority of the jobs are in the suburban and exurban rings.

Question the assumptions—however obvious may be their logic—and some large and foolish blunders can be avoided.

What conclusions can be reached from these blunders in conservation? The most significant is that "conservation" as a thing in isolation, possessing an intrinsic good, does not exist. Only common sense and economic judgment can serve as effective guides. Better information can help to communicate common economic sense, but arbitrary incentives and standards are more likely to mislead than to guide; they are somewhere between marginally useful and counterproductive to the workings of the energy marketplace.

## SYNFUELS—DEMONSTRATION VERSUS RESEARCH

Let us try another experiment. Read the four quotes that follow without looking at the references at the end of the chapter.

- An industry expert writes that oil shale constitutes "a reserve available for extraction whenever the demand and the price [of oil] shall become great enough to warrant the establishment of a new industry to supplement the supply of petroleum from the oil fields. This time is now at hand."[8]
- "Why won't the Synfuels Corporation work? The real problem may be technology."[9]
- "Oil-Sands Projects Draw New Interest"[10]
- "Don't Give Up on Synthetic Fuels"[11]

Now, reader, try to guess when each of the four statements was published. They span a period of sixty-six years. It is nearly impossible to guess which one was published in 1918 (the first) and which was published in 1984 (the last).

In the sixty-six years separating the headlines, interest in synfuels has increased; at times during the last fifteen years, it reached a nearly feverish pitch. Technologies have improved markedly, and the real market price of the competitive product (i.e., natural petroleum) has increased an order of magnitude. Still, the production level of synfuels remains nil, with no significant increase planned or in sight. We will now violate our own rule, which states that all predictions are fatuous, and make a prediction: There will be no significant production of synfuels during any of the twentieth century. (Note the not-so-subtle point that if our prediction is wrong, at least the validity of our rule will have been demonstrated.) By "significant production," we mean significant relative to world energy demand, that is, several million barrels per day. Our logic is as follows.

- Present technologies work. That is, *technology is not the problem.* The problem is that no known, emerging, or speculative technology can compete on cost with the price of world oil, or with any reasonable expectation of a future price of world oil.

- No new technologies contemplated at present would reduce costs enough to allow the resulting synfuels to compete.
- The perceived public policy benefits aren't nearly great enough to justify the necessary subsidy, whatever its form. That is, any possible benefits of an importing nation being insulated from the world oil market are perceived to be less than the difference between the technological cost of synfuels and the world oil price.
- World oil price is not expected to increase sufficiently to allow synfuels to compete.

## The Economics of Synfuels

Most of the cost of producing synfuels from oil shale or tar sands is in the removal of the hydrocarbons from the shale or sand. This cost is devoted to removal and replacement of overburden, mining of fresh and replacement of spent shale or sand, crushing and moving the shale or sand, and the heating and cooling of the shale or sand, which is required to liberate the hydrocarbons. Although a great deal of attention is paid to the problems of upgrading the raw hydrocarbon thus separated, and to transporting it to market, these issues pale before the cost of handling materials. For every pound of oil recovered, some ten pounds of inert rock or sand must be mined, crushed, handled, heated, recooled, and disposed of (who knows where). We estimated that the cost of mining and retorting oil shale to be almost five times the cost of upgrading the shale oil—once recovered—to syncrude (Ball, Barbera, and Weiss, 1979).

One of the reasons we have such low expectations of technological breakthroughs is the high cost of materials handling. The technology of materials handling is as mature as any technology in existence. It dates back literally to the beginning of civilization. The different "technologies"—those of TOSCO, Paraho, and Union—are primarily different ways of getting the heat into the crushed rock; but this is not the challenging issue.

Synfuels from coal, lignite, or peat do not have the high inherent handling cost associated with oil shale and tar sands. The former face economic barriers of a different sort. First, the feedstocks—coal, lignite, and peat—are themselves perfectly marketable industrial fuels. Therefore, their alternate use value is related to their BTU value, whereas the alternate use value of oil shale or tar sands is zero.

The second cost barrier associated with synfuels from coal, lignite, and peat is the very low hydrogen-to-carbon ratio of the feedstocks. When the feedstocks are burned directly, the low ratio is not a factor. But when the object is to manufacture from them a *hydro*carbon, then the hydrogen-to-carbon ratio becomes important. The problem also exists for oil recovered from shale (or tar sands), but less so; it is a much greater barrier in the case of coal (and lignite and peat), since the ratio in coal is much lower than it is for the hydrocarbons (kerogen) contained in the shale.

As a result of the low hydrogen-to-carbon ratio, huge quantities of hydrogen must be added to coal to upgrade it to any hydrocarbon "synfuel," whether liquid or gas. The ultimate commercial source of such hydrogen is water—itself a product of combustion. Therefore, one perspective on the manufacture of synfuels from coal is that it requires a reversal of the combustion process: the "unburning" of water. Any process whose objective is to produce a fuel is intrinsically inelegant thermodynamically; therefore, we are not surprised to find that the process has so far proved to be economically prohibitive.

Capital costs represent a large portion of the total cost of synfuels. The capital intensity of synfuels (i.e., the capital cost per barrel of daily production) is about ten times that of the most capital-intense crude oil production in hostile environments like the North Sea or Alaska's North Slope. In addition to this high capital intensity, the operation of synfuels plants has a very high labor component, and a short life for the materials-handling capital equipment (digging and moving equipment). These factors result not only in total costs much higher than world oil price but also in very high fixed and marginal (i.e., variable) operating costs.

The unpredictability of world oil price and the nascence of synfuels technology are often cited as the main barriers to the commercialization of synfuels (Stanfield, 1984). Instead, the opposite is the case. The expected world oil price levels and the maturity of synfuels technology are the major impediments.

Most of the "commercialization" projects were applied to mature technologies. Some true *research* was (properly) conducted on *new* synfuel technologies, such as in situ recovery (i.e., the underground recovery of hydrocarbons from oil shale or the underground conversion of coal to synfuels). As we discuss in a later chapter, it is not unusual to discover and develop a new technology for making a mature product, or for accomplishing a mature process or task. For example, as the price of fuels for internal combustion engines climbed, no serious consideration was given to commercialization projects applied to the mature technology of uneconomical external combustion automobile engines. But true *research* into *new* external combustion engine technology might have been appropriate.

With world oil prices expected—even on the high side—to remain low compared to synfuels costs, the maturity of synfuels technology mitigates the hope that their costs can be significantly reduced. The continued optimism of policymakers about the cost of these technologies has only served to cloud the facts.

## The Blunders

**A fallacious context.** The role of synfuels in the world energy picture has been complicated by a number of factors. Many stem from the naiveté of the energy-consuming nations in general, and the United States in particular, concerning the nature of world energy. Our understanding has increased significantly over the last dozen years, but the learning process has been slow and painful.

From the perspective of the late 1980s it is difficult to appreciate the emotional and romantic wave of concern about energy that swept the United States a decade ago and created the environment in which much so-called analysis was conducted. Appeals were made to the emotions, not to logic. In addition, opposition to some energy initiatives was political and social rather than economic. An example is the opposition to nuclear power, which focused on the entire nuclear fuel cycle—particularly waste disposal—and the power of the technobureaucracy that underpins the nuclear industry. Slogans tended to replace analysis. Vestiges remain even today: "Split Wood, Not Atoms," and "Solar Employs; Nuclear Destroys." Alternatives to imported oil became an intrinsic good, and synfuels became for many in power an article of faith. The question was not, Do synfuels make sense? but, Do you believe in synfuels? The implication was that not to do so was somehow unpatriotic.

The popular view, adopted by the U.S. government, was that we were going to run out of energy, and soon—in the next five or ten or fifteen or twenty years. After that, demand would exceed supply. The gap between the two would have to be filled by synfuels. Therefore, the huge future demand for synfuels was calculated by using simple projections that ignored all principles of microeconomics, not to mention the realities of the world oil market. Although this kind of analysis and policymaking have been largely abandoned, all U.S. synfuels projects—and, to a large degree, our confused way of thinking about them—remain as their legacy.

**The wrong focus.** Since the beginning, the U.S. government focus has been on demonstration; that is, the construction of full-size plants to demonstrate that synfuel production technologies would work, and to determine their cost. The intention was to demonstrate each of the more promising technologies. The assumption was that if the government provided artificial incentives (e.g., loan guarantees) for the demonstration plants, then the private sector would flock to replicate them, driven by nothing more sophisticated than the profit motive. The intention and assumption of the U.S. government's Synfuels Corporation did not change. The corporation was the last artifact of the failed national policy, except for a couple of very high cost synfuels demonstration plants that the Synfuels Corporation had built before it expired, which are still in existence.

The problem here is focus. The mature technologies are known to be too expensive to be competitive. The appropriate approach is not to demonstrate known technologies, but to invent new ones (Ball, 1981). This means *research*, not demonstration. Although research has been encouraged in other fields of alternate energy sources, like solar, it has not been encouraged nearly so much in synfuels, with the exception of in situ.

In situ is the process whereby the benefit of the energy resource is captured without excavating the resource itself. In the case of shale, this might involve separating the oil from the shale while it is still underground. The process is not yet technically feasible and progress to date has been quite disappointing. On an absolute basis, however, little work has been done and the potentials are great. In situ is one area of synfuels we know very little about. It requires improving our basic knowledge of the dynamics and control of subterranean thermodynamics, mechanics, and fluid flow. A breakthrough would not only apply to synfuels; perhaps more important, it would also apply to petroleum reservoir management, and could lead to significant improvements in the primary, secondary, and tertiary recovery of conventional petroleum.

An obvious advantage of in situ synfuels is the elimination of most or all of the solids handling. A less obvious advantage is that it would use, at least in part, fuel whose alternate use value is near-zero, such as coal before it has been mined, or kerogen before it has been extracted from the shale.

**Inadequate consideration of alternatives.** In the poorly conceived rush toward synfuels in 1974, several attractive alternatives were ignored or received only lip service. The alternative receiving the shortest shrift was the strategic petroleum reserve—the nation's inventory of produced petroleum. This alternative should have received much more serious attention, and could have been effectively and economically implemented on a much larger scale.

**Naive perspective.** The hopeful policy on synfuels was undergirded by a naive perspective that went something like this: We need lots of energy, and we have lots of oil shale, tar sands, and coal. Therefore, these abundant resources should provide the needed energy sources. Of course, the logic missed two crucial points. The first was economic. We might as well talk of the sea as a huge resource for gold, or even for potable water, or whatever. And the second is similar: need is a relative economic concept, not an absolute one.

**Flawed view of energy policy.** The United States and its leaders did not understand fully the limitations of the government in *directing* the energy sector. The adoption of an energy policy came to be viewed as a solution per se (Ball, 1977). This led to a number of absurdities, such statements as, "With the NEA [National Energy Act], we will save 2.5 to 3 million barrels a day by 1985, compared to what we would otherwise have required for an estimated balance of payments savings of approximately $14 billion in current dollars (as much as $20 billion in 1985 dollars)" (Schlesinger, 1979). With government a bit player in the energy economy, its saying so would not make it so. Government policies failed to answer the fundamental question: If private investors will not invest in synfuels, then why should the government invest their tax dollars for them? We did not fully appreciate the restrictions on the government's role in a market economy where the government is neither a significant producer nor consumer.

**The wrong industry.** The wrong industry has been promoting synfuels from oil shale and tar sands (Ball, 1979). As indicated above, the important technologies are related to handling solid materials, not to liquids processing. Therefore, the logical industries with relevant expertise would be the hard-rock mining and construction industries.

The petroleum industry is the best in the world at geophysical interpretation, the drilling of wells, reservoir management, and the transportation and refining of oil. But it brings little if anything to the table when it comes to mining, moving, heating, and cooling sand and rock. The recovery of oil from shale would be an assignment more appropriate to the coal, copper, or steel industry than the petroleum industry.

**A forgotten perspective.** At least two of the most popular technologies considered for synfuels originally had as their goal, not a low-cost synthetic fuel, but a novel way to reduce pollution from the combustion of coal. These two are Gulf's solvent refining of coal and the medium BTU gas produced from coal at the Coolwater integrated gasifier combined cycle (IGCC) plant. That a synthetic fuel is produced as an intermediary product is merely an artifact in both processes. They were intended primarily as alternatives to de-ashing and stack-gas scrubbing in the combustion of coal.

## Synfuels Projects

In its first three years, the Synfuels Corporation awarded two contracts totaling $740 million. Dow received $620 million for a 5,000 Bbl./day (BTU equivalent) plant to convert coal to medium BTU gas in Louisiana. The remaining $120 million was for the 4,300 Bbl./day (BTU equivalent), 100-megawatt Coolwater plant in California, for the conversion of coal to gas to electricity. Its owners are Bechtel, Electric Power Research Institute, General Electric, a Japanese consortium, Southern California Edison, and Texaco.

The two largest projects actually predate the formal incorporation of the Synfuels Corporation. The first is the Great Plains coal gasification plant in North Dakota. It is owned by Pacific Lighting Corporation, Tenneco, and Transco. It will produce 125 million standard cubic feet per day (mmscfd), or 21,000 Bbl./day BTU equivalent. The second is Union Oil's shale oil plant in Colorado. It has been experiencing operational difficulties, and allegedly violates environmental and worker safety and health standards (Stanfield, 1984).

In Canada, things look only slightly better. Sun's midcontinent tar sands operation continues to operate after twenty years. It is reported to be profitable. A decade after completion, Alberta's facilities for recovering syncrude from tar sands continue to operate. Ownership has changed often, and most recently is in the hands of Exxon, Amoco, Petro-Canada, and an eight-company consortium. Capacity is 170,000 Bbl./day. Announced spending programs, largely in situ, would increase the capacity to 300,000 Bbl./day. "Costs" are reported at $15/Bbl., with no more definition of costs than that they "exclude profit and taxes" (Martin, 1984).

## What We Should Do with Synfuels

Enthusiasm over synfuels these past dozen years rests upon an oversimplified assumption: since the price of natural crude or natural gas has gone up or, as some of the most naive have said, since we are running out, we need a source of *synthetic* crude or gas.

Such a perspective is not necessarily wrong, but it is limiting. Perhaps the point can best be made by analogy. The first automobiles were called horseless carriages, and that is exactly what they were. An engine had replaced the horse, but the body, frame, and chassis were still those of a carriage. It took several decades before an entirely new *system* designed around an engine and the passengers began to emerge. A similar example is the farm tractor, the wheel spacing of which is still determined by the space required between horses hitched to a team. Even today, we have not determined the optimum row spacing for crops, and then designed tractors to fit. We haven't developed a *system* of wheel spacing and spacing between the rows of crops that exploits engine-driven farm implements. A dozen years ago fundamental concerns over conventional crude and gas were raised by high prices, uncertain prices, uncertain supplies, and so forth. Regardless of the concern or what inspires it, only the most naive would confine their search to looking for direct replacements for conventional crude and gas. Yet, this is exactly the emphasis and appeal of synfuels. The true implications of the fundamental concerns are not, How do we synthesize crude? but rather, What are the alternative systems for obtaining the electricity, heat, and so forth we have conventionally derived from natural crude and its products? (See Ball, 1983.)

### Photovoltaics and Renewables

After the intial shocks of the 1973 energy crisis, the focus of the government of the United States and those of much of the Organization for Economic Cooperation and Development (OECD) was on supply substitution, i.e., on the development of energy supplies that were not under the thumb of foreign oil. From the perspective of this so-called security, the attractiveness of renewable resources was undeniable. Of the renewables, solar energy systems quickly became the most popular. The sun shines often, just going to waste if it isn't being collected and used.

In 1973, there were no significant government-supported energy research programs outside the nuclear field. Five years later the newly created Department of Energy (DOE) had a budget of $11 billion, a significant part of which was the carryover from the nuclear research days, including nuclear weapons research. The solar component of the budget was a mere $393 million, and peaked at $559 million in 1980 and $552 million in 1981, the end of the Carter administration. Figure 3-1 shows the Department of Energy pattern of expenditures, including those for renewables (included under R&D), in total dollars from 1978 to 1986.

Within the DOE, money for research and development of high-technology options such as photovoltaics (PV), wind, and ocean thermal energy conversion (OTEC) peaked at 20% of the total budget in 1980–1981 during the Carter and the Reagan administrations. During this period, research technologies in which there was felt to be the potential for a breakthrough were separated from the technologies in which only marginal changes were possible. The heating technologies fell into the latter category and included water heating, passive solar for buildings, and thermal storage. (See Figure 3-2.)

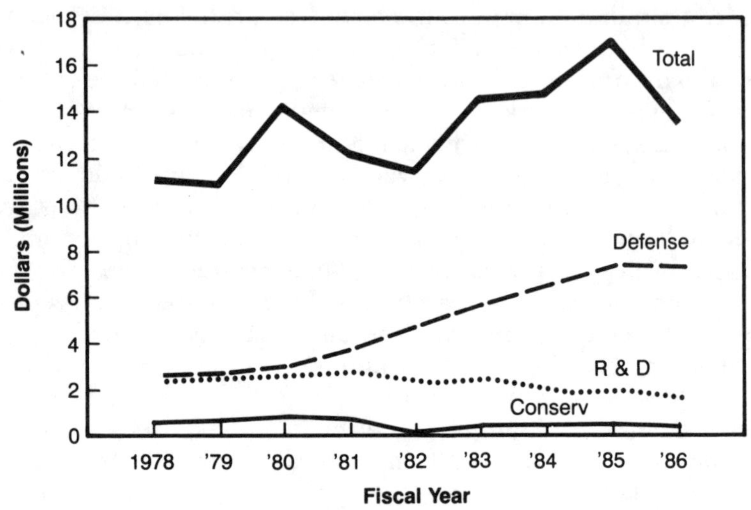


*Source:* Executive Office of the President, *Budget of the United States Government* for fiscal years 1978–1986.

**FIGURE 3-1 Department of Energy Budget.**

The real hope was in photovoltaics and solar thermal electric, each of which received no more than about 5.5% of the research budget in fiscal years 1980 and 1981—the high point in funding. The critical questions were, How fast could new technologies enter the market? How fast could the price per unit come down?

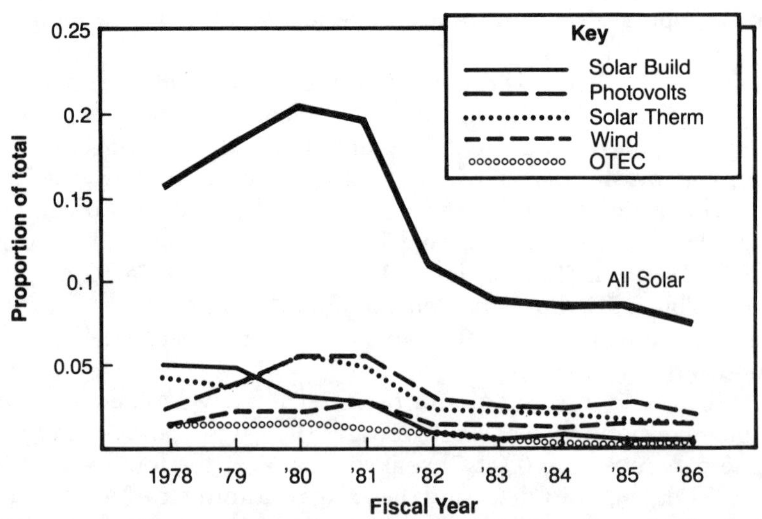


*Source:* Executive Office of the President, *Budget of the United States Government* for fiscal years 1978–1986.

**FIGURE 3-2 Department of Energy Solar Tech as Percentage of Energy R&D.**

In Chapter 5 we will describe the blunders associated with believing that the experience-curve function—the more you make, the less the per unit cost—can be adapted to all products in all phases of their life. Here we will look at how fuzzy thinking about the relationship between increase in production and decrease in cost can lead to unrealistic expectations about technology development.

Prior to the oil price shift of 1973, PV was a technology used exclusively to provide power for space stations. By paying between $50 and $100 per peak watt ($W_p$), NASA was able to supply nearly 100% reliable power to space applications.[12] Space, with a cloudless sky twenty-four hours per day, provides a perfect application for PV, and cost was never an issue at NASA—only reliability. Some limited commercial applications of photovoltaics had developed in which intermittent power was acceptable and where, again, price was not an issue. The classic examples are cathodic protection of pipelines and Coast Guard buoys. In 1973, that was the entire market for photovoltaics.

What occurred after 1973 in PV development was not dramatically different from what occurred with other technologies: it joined the horse race to reach costs competitive with escalating energy prices. All of the solar and renewable electric systems and the exotic fuels systems set off in competition with one another to reach the major goal of producing electricity at prices that matched those of a conventional power plant. Although on the surface the competition was to produce inexpensive electricity, in fact the competition was for federal funds. The better the team working on the problem could sell its ideas, the more federal funding was received to proceed with the research.

Because PV was a space technology, the initial players were from NASA. Nearly all NASA laboratories had a piece of the early pie, as did a number of corporate space contractors. By 1976, the original set of NASA laboratories and contractors had been sorted out, with the effective groups remaining and the less effective ones dropping by the wayside. At the same time, new groups entered the picture. The Jet Propulsion Laboratory at Cal Tech became the lead center; the Lincoln and Energy laboratories at MIT became prime contractors, as did Aerospace Engineering, NASA Lewis, MERADCOM of the Army, and Sandia Laboratories (first of AEC and later of DOE). The result was a research and development group experienced in competing for federal research and development funds.

Where did the PV program succeed and where did it fail; and, could it really have met its stated goals? These are the key questions.

The original PV program goals were developed at meetings held in Cherry Hill, New Jersey, in October, 1973 (Cherry Hill, 1973). At these meetings two goals were set for the program. (See

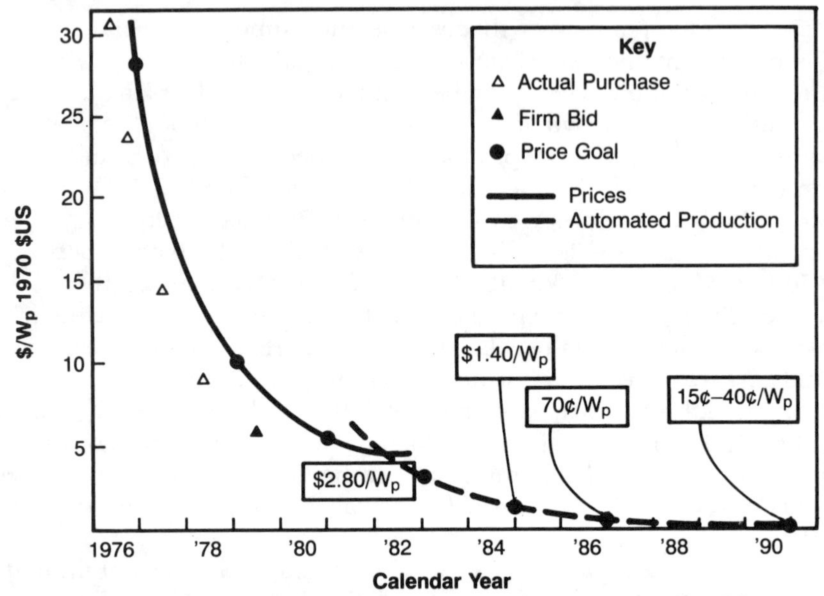


*Source:* Adapted from DOE Photovoltaic Program Planning Documents, Washington, DC.

**FIGURE 3-3 Department of Energy Photovoltaic Price Goals.**

Figures 3-3 and 3-4.) Figure 3-3 shows the goals in terms of price versus time. (Note that this is not an experience curve, which would be price or cost versus quantity.) Figure 3-4 shows a quantity versus time curve. Early discussions of the technology included a poorly defined experience curve, but it was quickly replaced by early versions of what appears in Figure 3-3, which became synonymous with the goals of the photovoltaic program.

The goals of the program were to achieve price reduction through increased volume of production as shown in Figure 3-3. The federal government was to be the consumer and the size of the purchases and the price were to be fixed as functions of time. The triangles in Figure 3-3 represent the timing and the price of the purchases. Each purchase shown was to be roughly double, in $kW_p$, that of the previous purchase. While neither well documented nor well articulated, the underlying assumption was that we could, in some sense, force the technology down this surrogate experience curve.

It was felt that government purchases could substitute for the private market. The government could purchase preset quantities of photovoltaic panels on a schedule where manufacturers knew the quantity the government was actually going to buy and at what price. Both price and quantity were to be determined by their location on the surrogate experience curve for the technology.

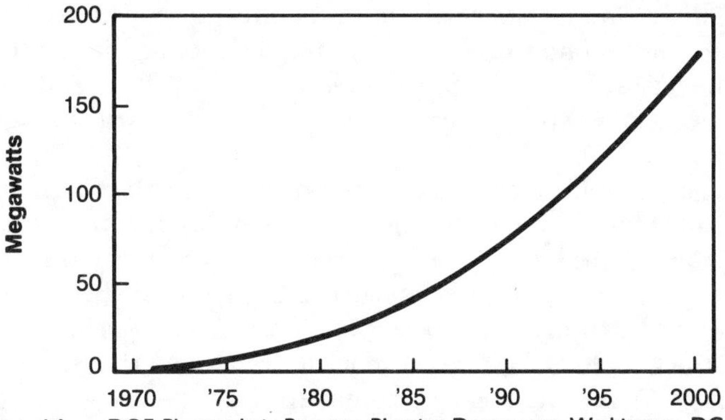


*Source:* Adapted from DOE Photovoltaic Program Planning Documents, Washington, DC.

**FIGURE 3-4 Photovoltaic Production Goals (megawatts annually).**

What were the blunders? There were three interrelated blunders associated with not understanding the assumptions upon which the experience curves were based—the foundations of the photovoltaic program goals.

The first was that the relationship between Figures 3-3 and 3-4 was not questioned until 1977–1978. Both curves were seen as possible, both curves were set as goals, and both were *thought to be internally consistent.* They were not. Underlying both curves should be an experience curve which could relate the unit cost (not price) of production to the cumulative quantity produced—not time—*in a competitive market.* The experience curve would provide the unit *cost* information upon which the manufacturer could base a price. The experience curve could also provide a portion of the data required to generate a projection of quantity to be sold as a function of time (see Figure 3-4). Figures 3-3 and 3-4 *should* have been related but they were not.

To the Energy Research and Development Administration (ERDA) and then the DOE, the program goal was to provide a significant contribution (gap filling) to U.S. energy needs by the turn of the century. The program was not initially perceived to have the goal of creating a market, yet that was what the government's role became as it effectively dictated price, quantity, and time. The major portion of this blunder was assuming that it was possible, in effect, to dictate the supply-demand relationship in advance and that by having the government establish the market through forced, prestated quantity purchases, it would be possible to drive the price of the technology down.

The second blunder was that the solid curve, which represented the forecasted economics of the existing technology, indicated that it would never reach an acceptable cost level. Despite this, the decision was to proceed to spend money, hoping to walk down the experience curve.

The third blunder was an attempt to justify the second blunder. Figure 3-3 shows two nearly smooth price reduction curves as a function of time. These are actually curves for two highly differentiated technologies. Reduction in cost at the high end of the curve was to be achieved by improved production methods on *existing* technology, that is, the Czochralski ingot. The final price reduction to achieve the goal of $0.40 per $W_p$ was to come from a new technology, either edge-defined ribbon growth or amorphous technology, both of which were on the drawing board but had not been proven viable, much less economic. In the 1970s, there was absolutely no basis upon which to forecast the production behavior of the second technology, since it had yet to come off the drawing board. Thus, while conceptually a second, related technology *may* substitute for an earlier one, projection of a relationship requires information, information that did not then exist.

We have used photovoltaics as our example because one of the authors was a part of the program. The blunders in the photovoltaic program were repeated with each of the major alternatives in the race for funds and contribution to the U.S. energy needs. The wind program focused on larger and larger turbines, hoping to capture economies of scale. These, in practice, turned into *dis*economies. Wind farms with many small machines finally proved to be more economic. Research on ocean thermal energy conversion systems (OTEC) focused on increasing scale as well, looking for economies in thermal gradients of ocean water. Even though engineering scale improved the economics on paper, it was biological *scales* (predominantly barnacles) that built up on the first-stage heat exchangers that brought the system down. The race was never won or lost, merely slowed after the third lap. Of the renewable technologies shown in Figure 3-2, photovoltaics continues to have the largest budget. Investment in photovoltaic *research* is likely to be the area of largest payoff in the decade ahead.

So, what have we learned? We have come to recognize that descriptors of market mechanisms are useful in real markets, but government purchases are not a substitute for the private market. And the experience-curve phenomenon, when used correctly, can be a fair description of the relationship between increase in the total number of units produced and cost per unit. Used incorrectly, it leads to very poor policy decisions.

What would the analysis have looked like had we done it correctly? The first step of the analysis would have been to develop as thorough an understanding as possible of the experience curve for photovoltaics, bearing in mind that the experience curve reflects costs of production, not market prices. The second step would have been to forecast the demand for photovoltaics as a function of price (elasticity), of price of alternatives (cross elasticity), and of time. With these data it would have been possible to begin development of a realistic set of forecasts and goals for the experience-curve-based costs of production. Done this way, it is unlikely that there would have been the heavy emphasis on single-crystal silicon technology development. The concentration, in all likelihood, would have been on greater research on the technologies required for meeting a large-scale, commercial market.

## GAS: FUEL OF CHOICE?

Natural gas is a unique hydrocarbon.[13] Chemically it is methane ($CH_4$), the simplest possible hydrocarbon. Until recently only a limited amount of exploration has been carried out explicitly for gas because it was generally a by-product of the exploration for oil.

The combustion of natural gas is simple and clean, compared with that of the more complex hydrocarbons like gasoline, home heating oil, and residual fuel. It is, or can easily be made to be, free of sulfur contaminants, although it does produce oxides of nitrogen when burned (the quantity formed is largely a function of the temperature of combustion). As a result, it is a clean fuel. Gas arrives underground to the site (liquefied natural gas aside), and it requires no significant refining before use. It comes in its own container, best stored in the producing well itself, or possibly in large underground caverns or salt domes. In today's U.S. market, roughly 30% of gas goes to residential, 15% to commercial, 35% to industry, and 20% to electric utilities. The prices of interstate transactions between the wellhead, pipeline, and distributors have been federally controlled since the 1950s. Gas prices to the consumers have been continuously regulated by state commissions.

The market for gas from the late 1940s to the late 1960s required only minimal thought and planning. Demand would expand about as quickly as new hookups could be made. At the then-current regulated prices, the system had enough gas for supplies. The system worked effectively through much of the 1960s until federal price ceilings for gas at the well constrained the entry of gas to the pipelines and therefore to the final consumer. Essentially, the regulation reduced the ability of distribution companies to continue expansion through new hookups. (See Figure 3-5.).

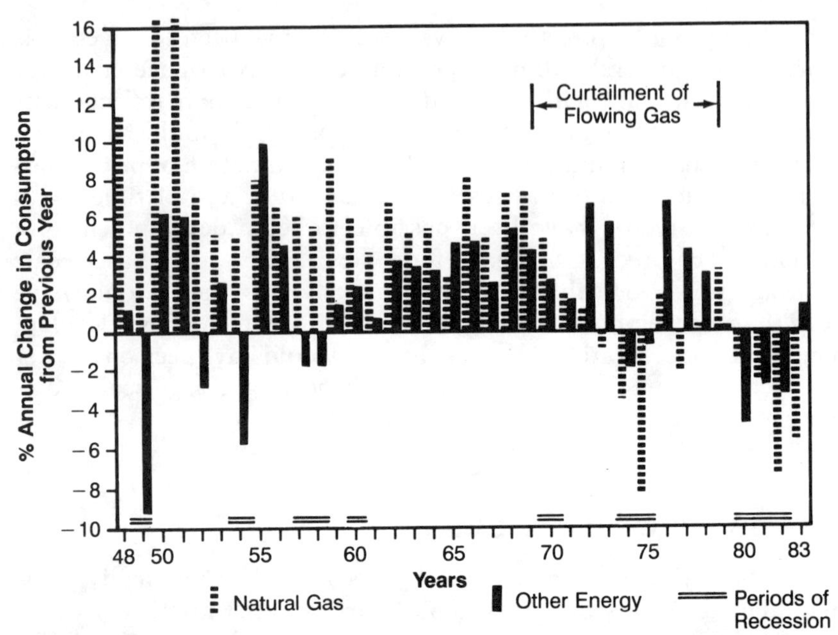


*Source:* Paul R. Carpenter, Henry D. Jacoby, and Arthur W. Wright. 1986. "Adapting to Change in Natural Gas Markets," MIT Energy Laboratory, Working paper number MIT-EL-85-007WP.

**FIGURE 3-5 U.S. Growth Rates for Natural Gas and Other Energy.**

It is not difficult to see the psychological ties with gas over the years. Traditionally, consumers of gas use it as their principal source of energy. Liquid oil can be transported around the world by an almost unlimited number of routes, and by a mixture of almost any conceivable carriers: pipeline, ship, trucks, rail, drums, and so forth. But a gas customer is bound to the pipeline system in place; it connects him physically and directly to the producing wells. No alternative sources or means of transportation can be substituted, except in the long term and after huge capital expenditures. Flexibility is, therefore, limited. As a result, shortages, however caused, raise anxieties to the point of panic for both residential consumers and regulators. Consumers have difficulty in seeing the difference between a regulatory (price) induced shortage and a shortage in capacity (pipeline, for instance). For them, a frosty December night without heat reduces the difference to pure rhetoric.

During the 1973–1974 oil price shock, interstate gas prices did not adjust sufficiently rapidly to reflect the market position of gas. The result was chaotic.

> Increases in federal price ceilings could not prevent the situation from deteriorating even more (relative to the price induced shortages of the late 1960s). In 1977 the Carter administration proposed overhauling the whole system, and after an 18-month legislative battle there emerged the Natural Gas Policy Act (NGPA) of 1978. This bill provided for the phased decontrol of the prices of new gas, with removal of controls on January 1, 1985; perpetual control of old gas prices; and extension of the system of federal ceilings to intrastate field markets.
>
> The NGPA decontrol process had only just begun when the second oil shock pitched the industry into what became known as the market ordering problem[14]—renewed shortages, permissible field prices ranging from $1.00 to $11.00 per thousand cubic feet (Mcf), and fears that a "fly-up" of new-gas prices on January 1, 1985, would create chaos. Contract terms responded to these confused conditions: higher gas prices were negotiated wherever they were permitted. Where prices were fixed, substitutes for price (like "indefinite" price escalators and higher "take-or-pay" requirements) were introduced. Unfortunately, the changes rendered the system even more vulnerable to the collapse that was to come.
>
> And come it did, in the form of the recession of 1981–83 and the downward slide in oil prices beginning in 1982. Structural shifts in U.S. manufacturing[15] and a warm winter in 1982–83 added to the problems of the gas industry. The demand for natural gas fell sharply. But the gas industry lacked the flexibility to respond; instead of prices dropping to clear markets, they continued to rise.[16] Attempts to sell gas more competitively ran afoul of protests from firms whose markets were threatened. *The gas industry had begun life on the demand curve, and an industry-wide crisis was at hand.* (Carpenter, Jacoby, and Wright, 1986, p. 16.)[17]

The Fuel Use Act constrained heavily the use of natural gas in new applications. Gas was specifically not allowed to be used in boilers because, we believed, that "the demand for gas [was] considerably higher than the amount that can be supplied" (National Energy Plan, 1977). Gas became an energy source with a moral value—it became "too valuable to burn" in boilers. It took the creativity of the marketplace to find ways around the regulations and to finally bring down the "too good to burn" attitude of the 1970s. In 1987 we finally saw the effective end of the Fuel Use Act provision.

The most striking example can be seen in the electric power industry where gas-based technologies became increasingly attractive both because of fuel price and because of the modularity of the technologies themselves. To get around the regulations that forbade utilities to use natural gas in new boilers, the utilities arrived at a whole set of creative ways of circumventing the law. One was the cogeneration option. As long as there was some cogeneration, it was legally possible to burn gas. How little cogeneration, and how economic the cogeneration component, were open questions.

A second option was even more creative: staged development. Coal and coal gasification technologies were highly acceptable. EPRI and Southern California Edison had demonstrated at Coolwater, California, the integrated gasifier combined cycle system as a technically viable (though not economically attractive) alternative. Although you could not get a license to build a natural gas fired combined cycle plant, you could get a license to build a phased IGCC. The first two phases were fueled with natural gas. The final phase would add the coal gasification system. Because the coal gasification system accounts for more than 50% of the cost of the IGCC, it is unlikely in most instances that this phase would ever be reached. In effect, the IGCC component was added to the plans in order to obtain an operating license.

The gas situation has changed. Deregulation of the markets, including both producer and pipeline, is occurring, and supplies have reentered the market in abundant quantities. Surplus production capacity abounds, with no shortage in sight in the view of many observers. New supplies have been found and are being marketed to the United States from both Mexico and Canada. The result has been referred to as the gas bubble. Neither the shortage nor bubble image reflects the reality that there is today a more mature energy marketplace that includes natural gas as a major player.

Gas, like all other energy supplies, is no more than a means to an end. Its use is a function of economics, acknowledging that existing capital stocks are a significant part of the overall economic picture. We left the 1970s with a shortage mentality. The question is whether we will leave the 1980s with a more rational attitude toward natural gas. Natural gas, it is suggested, *may offer* one set of lessons for other energy sources—specifically electricity—as we move ahead. That is, gas pipelines and gas contracts have been deregulated, or at minimum allowed to move more freely in the market. Gas pipelines are now essentially common carriers. Distributors and even consumers are now able to strike any deal they can at the wellhead and obtain, subject to some constraints, access to the pipeline system to deliver it to the point of final demand. Suppliers are competing to sell gas, pipelines are moving the gas, and distribution companies are marketing their product. The system has some checks and balances.

Despite these signals, a significant number of people remain unconvinced that natural gas is available in adequate quantity. They are unwilling to admit that a natural equilibrium between supply and demand for natural gas can be driven by unregulated market pricing. The regulators are still trying to figure out what, if anything, needs to be regulated. Despite this, the system is working. Shortages, though still talked about, have moved into the background. The issue most of the gas industry seems to be facing today is, how to make money in the face of competition.

## NUCLEAR

The development of nuclear power in the United States is an outstanding example of what *not* to do following achievement of unquestionable scientific and technological leadership in a critical field. Although nuclear energy is an integral part of the energy picture, its progress from 1945 until 1975 happened *outside* the first two energy phases described in Chapter 2. Nuclear energy was a by-product of the military effort of World War II. During the thirty years following the end of the war, the development of nuclear power was driven not so much by market forces as by research, development, technological, and political motives.

The development of the atomic bomb in the United States was an unprecedented scientific success. It translated a theoretical idea into a practical operating device by engaging the most prominent scientists and engineers in an enterprise of the highest national priority. Under the exigencies of war, the task was accomplished in a very short time. The success indisputably made the United States the world leader in all facets of nuclear energy, and it provided us with the opportunity to be the first to develop nuclear power for nonmilitary applications. This enormous success, however, also led to blunders that may prove devastating if not fatal to our peaceful use of atomic energy in subsequent years.

Development of peaceful nuclear power did not begin immediately after World War II. On January 14, 1946, Bernard Baruch presented a U.S. proposal to the United Nations Atomic Energy Commission, which would have established an international agency with exclusive authority to own all nuclear materials and to conduct all dangerous nuclear operations (Scheinman, 1985). The proposal was not accepted, and the U.S. government decided that the science and technology developed in the Manhattan Project would remain secret. Congress passed the Atomic Energy Act of 1946, which prohibited any peaceful nuclear cooperation with other nations until Congress was satisfied that international safeguards were in place. The policy of secrecy not only inhibited commercial development of nuclear power in the United States, but proved ineffective in preventing other countries from developing nuclear energy themselves—first as bombs, then as nuclear power. The Soviet Union conducted tests of a nuclear fission device in 1949, and of a thermonuclear device in 1953. Britain conducted a test of a nuclear explosive in 1952.

The failure of U.S. policy to prevent the proliferation of nuclear technology led to a dramatically different approach. In a speech to the United Nations on December 8, 1953, President Eisenhower outlined the Atoms for Peace program, under which the United States would declassify the information necessary for development of the peaceful use of atomic energy.

The new policy provided the basis for an accelerated development of nuclear power beginning in 1954, not only in the United States, but also in other countries. The technological leadership already amassed by the United States, however, remained overwhelming. Export business in nuclear power for a while was a significant factor in the U.S. balance of trade after World War II. By contrast, the nuclear option today is almost dead in the United States, and the performance of nuclear reactors in other countries—Japan, France, Germany, and Finland, with Finland using a Soviet design—surpassed those in the United States. What happened?

In the early years, numerous nuclear reactor technologies were available for investigation: water-moderated homogeneous reactors, molten salt reactors, organic cooled and moderated reactors, sodium-cooled and graphite-moderated reactors, and carbon dioxide ($CO_2$) cooled and graphite-moderated reactors, to name a few. But these technologies became casualties of the U.S. atomic power program. Commitment to the light-water reactors was made long before studies on the other reactor concepts were carried out. This early commitment had its origin in the military atomic program. In August 1950, then Capt. Hyman Rickover concluded that a proposal originated by the Oak Ridge National Laboratories and taken up by Westinghouse was the best prospect for the development of a power reactor for submarines. Rickover chose on the basis favored by Robert Watson-Watt: Always select the third best. The first best never comes, and the second best comes too late (Perry et al., 1977).

The implications of Rickover's decision were far-reaching. The other U.S. electric giant, GE, decided to compete against Westinghouse with a modification of the light-water technology, and introduced the boiling-water technology. Even with the superiority of GE's marketing power, it found competition tough against Westinghouse's pressurized water reactors. The reason was that many customer organizations were staffed by civilian nuclear engineers who had retired from the military, where they had been trained in pressurized water reactors by Rickover.

Let us examine some of the U.S. government policies and decisions that, in part, undermined successful nuclear efforts in the United States. In January 1955, the Atomic Energy Commission (AEC) announced the first round of its Power Reactor Demonstration Program (PRDP), intended to bring private resources into the development of nuclear reactors. Three kinds of assistance were proposed. These were fuel-use charges, part of the preconstruction R&D, and R&D of post-construction operations. Under the terms of the first round, the AEC accepted two private reactor projects for federal assistance, the Yankee and the Fermi, and approved a third, the Hallam sodium graphite reactor. The Yankee pressurized water reactor (PWR) was a striking success as an operating plant, principally because of excellent management, supported by a consortium of a dozen or more utilities and manufacturers. Moreover, the PWR technology was advanced enough for demonstration. By contrast, Fermi's failure reflected the immaturity of breeder technology and unwarranted optimism about the resolution of its technological uncertainties. Hallam was a project of a Nebraska company that agreed to provide only 25% of the estimated total costs.

The second round of the PRDP was intended to attract the participation of small, publicly owned utilities in the construction of small experimental reactors. It only became obvious later that small utilities were not good candidates for the technologies, which were still highly uncertain. It was a well-known fact that U.S. utilities, except the largest ones such as American Electric Power, very seldom conduct R&D on their own. How this fact escaped the attention of the AEC is not now understandable. For example, a Rand report stated:

> In retrospect, the reasons for the failure of the second round projects became obvious. The AEC has coupled reactors of high technical risk with low sponsor capability to assume the associated financial risks. If inducing small utilities to participate in the demonstration program were important to the AEC, low risk reactors of conservative design, probably small light water reactors, would have been built. . . . The institutional and economic setting of the second round of the PRDP was incompatible with the technology it invoked.

> By means of such demonstration projects, the AEC had hoped to determine whether several varieties of nuclear reactors had commercial potential. But in the process, the important distinction between demonstration projects and experimental R&D projects was blurred, and most second and third round PRDP projects lived in a half-world between success and failure. (Perry et al., 1977, p. 13.)

We could go on reviewing the past management of the PRDP activities, but they are only part of the story. The third-round situation was not very different from the second. Let's look at some other aspects.

In the early years, the AEC had a dual role: promotion of nuclear power as an attractive option for electricity generation and regulation of safety. Of the two, AEC was clearly more enamored with the first. The regulatory part of the AEC never enjoyed the same status within the commission as the promotional part. The regulatory part was small, with little say in policy and budgets. A result was limited emphasis on safety R&D. The belief was that if the economic attractiveness were demonstrated, the technical problems would take care of themselves. The current status of nuclear waste disposal is one obvious by-product of this philosophy.

The same spirit was adopted by the manufacturing sector. The giants in the sector believed that if the volume of business expanded rapidly, the economic attractiveness would prove itself by the well-known learning-curve effect and the distribution of R&D costs over a large base. All manufacturers wanted to gain a strong foothold in the new and growing industry and the approach they followed was the traditional market-share strategy. To implement the strategy, they adopted another well-known concept: economy of scale. They offered bigger and bigger plants. In the period between 1960 and 1975, the average size of reactors offered increased at a rate of about 60 megawatts per year. Because of the time lag between first offering and operation, when the industry was offering and designing 1,300-megawatt reactors in the early 1970s, the largest reactor running was about 250 megawatts. Thus, each new plant designed was larger than the last, but the design was drawn before benefit had been gained in operating experience with the preceding size plant. As a result, there was no ride down the experience curve, since learning comes from making the same (or almost the same but with refinement) thing over and over again. In this setting, economies of scale are also largely inoperable.

The blind belief in the market-share theory, without due attention to its known warnings and limitations, went much further. Manufacturers offered many additional inducements in an effort to attract new orders: low-priced uranium from Westinghouse (a billion-dollar fiasco), and fuel-cycle service contracts offered by suppliers guaranteeing the performance of nuclear fuel and the reprocessing of spent fuel into a mixture of plutonium and other metallic oxides. The latter could have been another billion-dollar fiasco, except for President Carter's ban on the use of plutonium in the United States.

We cannot leave the subject of nuclear blunders without addressing the regulatory issue, particularly safety regulations. The AEC safety program was burdened with deficiencies, of which we will mention only a few. First, the manufacturers' market-share strategy forced rapid changes in reactor design. The utility industry was itself part of the problem: everyone wanted something different, and thus amplified and fragmented nuclear construction procedures by separating architect and engineering firms from manufacturers and construction contractors. This state of affairs effectively precluded any orderly development of licensing criteria. The lack of such criteria led to lengthy and expensive delays in licensing. Second, the AEC safety program was totally inadequate in scope, and was unresponsive to regulatory needs. The regulatory division, because of the way the AEC was organized, had very little control over the safety program. Third, there seemed to be no strong belief in the need for a safety research program. This belief can be seen in the strategy of simultaneous development, testing, and commercial deployment of nuclear reactors. Therefore, operating power plants became de facto testing laboratories.

Again, we quote from the same Rand report:

> By the mid-1970s, four flaws in the process of commercializing light water reactors were retrospectively being attributed to the earlier actions or inactions of the Atomic Energy Commission. The first was that regulatory and licensing delays had slowed the completion of individual nuclear plants. . . . A second was the questionable adequacy of the safety standards. . . . Third, changes in AEC requirements, sometimes retroactive, were alleged to have substantially increased the expense and uncertainty of reactor construction. And fourth, the inattention of the Atomic Energy Commission to public demands for more comprehensive safety and environmental provisions was credited with causing a decline in public

confidence in the AEC and the nuclear reactor industry. (Perry et al., 1977, p. 13.)

Looking back today, we can also see the mistakes in overlooking the back end of the fuel cycle, such as high-level nuclear waste and the ultimate decommissioning of aging plants. These issues now haunt us.

We must add that the AEC was hardly the only one to take missteps. The manufacturers played their part. Public confidence in the scientific accomplishments and capabilities of the United States was so overwhelming that no one doubted the ability of industry to perform the technical job at hand. It was appreciated that we must know enough about the neutron physics in the reactor to make sure that the nuclear portion of the system behaved properly. Complex and sophisticated computer programs were developed for that purpose. But not much attention was given to the balance of the plant. It was assumed to be just another power plant. Then, in the mid-1970s, the industry was plagued with component failures: valves, pumps, and stress corrosion problems. Even worse, the century-old problem of water chemistry was almost entirely overlooked. The industry seemed to have forgotten that a nuclear power plant is also a mechanical and a chemical plant as well. The ancient art of designing and handling steam-powered systems seemed to have been lost. It is significant that in the late 1970s one supplier found it necessary to build a multimillion-dollar valve-testing facility and an equally impressive pipe-testing facility. It is also significant that the question of the replacement of steam generators in a PWR plant was never thought through. For example, when maintenance was required on the steam system, no provisions had been made for accessing it, requiring that the containment wall be cut. This is equivalent to designing an automobile that must have the engine removed in order to change the spark plugs.

The industry seemed to have forgotten that a nuclear plant is a system—a system that requires an interdisciplinary perspective. A failure in any one of the more mundane disciplines can be as disastrous as the failure of the most glamorous one. It pays to remember that the glamorous discipline usually receives enough attention, and is well looked after. Problems in the field are usually in areas that the engineers and scientists tend to take for granted.

We have made no attempt to estimate the financial losses caused by nuclear blunders. It is safe to say that the magnitude is in the tens of billions of dollars. Of even greater importance, however, the blunders may have caused irreversible damage to two important U.S. industries. The electric power utility companies experienced a shock to their financial viability and a loss of public trust. The investor's maxim that electric utilities "never lost money and never went broke" had run into a disturbing new reality. And the U.S. nuclear industry, which had started the game with all the best cards, had lost its standing in the competitive global market.

## NOTES

1. For differing but more comprehensive views on the history of petroleum, we recommend Blair (1978) and Sampson (1975). Both works are well documented, and are worthy of attention, despite the patently prejudiced view both authors take of the petroleum industry. For example, they berate the petroleum industry for exploitation and price fixing, but they simply ignore the fact that when the oil companies lost control of production levels in 1973, prices went steadily upward for a decade! If one can rise above that kind of one-sided interpretation, the reader will find their story of the world petroleum industry a fascinating one.

2. Even British law provides that natural resources belong to the crown, under the premise that all things that migrate—water, waterfowl, deer, underground streams, and oil—belong to all the people; the crown or national government holds title in the name of the people. Remember, Robin Hood was guilty of killing the *king*'s deer. Even today in the United States, we still must buy a *federal* duck stamp to hunt migratory water fowl.

3. Unitizing fields is a relatively recent concept, and refers to viewing an entire field as a unit, as if it were jointly owned by the relevant mineral rights owners. The concept permits the wells to be drilled on a geologically and technically rational basis, with the *production* to be shared on an equitable basis by the owners of the mineral rights, which derive from the surface rights. The result is fewer wells for the same production, and therefore a lower cost for oil. Other countries have had this advantage all along.

4. It is a little-appreciated fact that second only to the Soviet Union, the United States is by far the largest producer of crude in the world. Its production in 1986 was over 50% greater than that of the next largest—Saudi Arabia. But U.S. production requires over a thousand times as many wells.

5. The price controls on liquid petroleum began when general price controls were mandated in 1971 (Grayson, 1974), and price controls were maintained on petroleum after those on all other products were discontinued. Price controls on crude oil continued until 1981, when they were replaced with the so-called excess profits tax (which is an excise tax, not a tax on profits). Gas price controls continue in effect as this is written (1988).

6. PL 94-163 was enacted on December 17, 1975, and subsequently amended by PL 94-258 (April 5, 1976), PL 94-385 (August 14, 1976), PL 95-70 (July 21, 1977), PL 95-619, the National Energy Conservation Policy Act (November 9, 1978), PL 95-620 (November 9, 1978), PL 96-94 (October 31, 1979), PL 96-102 (November 5, 1979), and PL 96-133 (November 30, 1979). The National Energy Act included PL 96-617; 618; 619; 620 and 621. These and subsequent amendments cover all aspects of U.S. domestic energy policy, including conservation.

In addition, the Energy Tax Act of 1978 (PL 95-618), signed by President Carter on November 9, 1978, provided specifically for residential tax credits.

7. The conclusions presented in this section are based on a research project carried out by R. D. Tabors and P. R. Rogers (Tabor and Rogers, 1980). The conclusions were unacceptable to the sponsor and therefore the study was not published.

8. Mitchell, 1918.

9. Stanfield, 1984.

10. Headline in the *New York Times*, May 28, 1984.

11. Headline in *BusinessWeek*, May 28, 1984.

12. A peak watt ($W_p$) is the unit of measure for photovoltaic power production units. It is the output of any photovoltaic generator based on incoming solar insolation of $1 kW/m^2$ at standard temperature and atmospheric pressure. Thus, a 10% efficient photovoltaic module of $1 m^2$ would be rated at 100 $W_p$.

13. It is interesting to point out that both in the United States and in Europe the "gaslight period" of the 1890s was based on gasified coal—"syngas," in today's vocabulary. This is a low to medium BTU gas made up predominantly of carbon monoxide and hydrogen.

14. The phrase "market ordering problems" refers to times at which the price of the product is dramatically out of sync with the price of products for which it is a substitute. In this case, primarily fuel oil.

15. As an example, the steel industry underwent a basic structural change. Demand for steel declined as lighter and more durable products were substituted in basic manufacturing. Automobiles used more plastics and alloys, for instance. The effect on the industry was to alter fundamentally the interindustry relationship of steel in the U.S. economy.

16. During the 1970s and into 1980 and 1981, we thought we were "running out of gas." Therefore, pipeline companies thought no commitment was too great to ensure "access," and producers had no trouble extracting brutal take-or-pay clauses. When the (inevitable?) glut followed, with its concomitant curtailment, it was less costly for pipelines to reduce their take of gas from older, lower-priced contracts with no take-or-pay penalty than to reduce their take from the higher-priced contracts, with their take-or-pay provisions. Therefore, as demand fell, the average price of pipeline gas actually *rose*.

17. Emphasis added. The gas industry is generally felt to have been demand driven. It was considered a competitive and generally superior fuel wherever it was available. In addition, its price was regulated. The idea that there was a supply and demand balance in a competitive marketplace was not yet recognized by either suppliers or consumers.

## REFERENCES

Adelman, M. A., and H. D. Jacoby. 1978. "Oil Gaps, Prices and Economic Growth." MIT World Oil Project. Working paper number (May) MIT-EL-78-008WP. Cambridge, MA.

Ball, Ben C., Jr. 1977. "Energy: Policymaking in a New Reality." *Technology Review* 61 (October–November), pp. 48–51.

———. 1979. "New Challenges to Management in the Synfuels Revolution." *Technology Review* 61 (August–September), pp. 34–35.

———. 1981. "An Investigation into the Potential Economics of Large-Scale Shale Oil Production." In *Oil Shale, Tar Sands and Related Materials,* edited by H. C. Stauffer. American Chemical Society Symposium Series 163. Washington, DC: ACS Books.

———. 1983. "The Future of Energy Technology." In *Global Technological Change: A Strategic Assessment.* Cambridge, MA: MIT Industrial Liaison Program.

———, and R. Barbera, and M. Weiss. 1979. "Shale Oil: Potential Economics of Large-Scale Production, Preliminary Phase." MIT Energy Laboratory. Working paper number (June) MIT-EL-79-012WP. Cambridge, MA.

Blair, John M. 1978. *The Control of Oil.* New York: Vintage Books.

Carpenter, P. R., H. D. Jacoby, and A. W. Wright. 1986. "Adapting to Change in Natural Gas Markets." MIT Energy Laboratory. Working paper number (January) MIT-EL-85-007WP. Cambridge, MA.

Cherry Hill. 1973. *Proceedings of the Workshop on Photovoltaic Conversion of Solar Energy for Terrestrial Applications.* NSF-RA-N-74-013, Cherry Hill, NJ, October 23–25.

DeGolyer and MacNaughton. 1987. *Twentieth Century Petroleum Statistics.* Dallas, TX.

Grayson, C. Jackson, Jr. 1974. *Confessions of a Price Controller.* Homewood, IL: Dow Jones-Irwin.

Jacoby, H. D. 1979. "The Oil Price 'Ratchet' and U.S. Energy Policy." *Kokusai Shigen* (*International Resources*).

Martin, Douglas. 1984. "New Interest in Canada for Oil-Sands Projects." *New York Times,* May 28.

Mitchell, Guy Elliott. 1918. "U.S. Geological Survey." *National Geographic.*

National Energy Plan. 1977. Washington, DC: Office of the President, Energy Policy and Planning.

Perry, Robert, et al. 1977. "Development and Commercialization of the Light Water Reactor, 1946–1976." Rand R-2180-NSF. Santa Monica, CA.

Sampson, Anthony. 1975. *The Seven Sisters.* New York: Viking.

Scheinman, L. 1985. "The Non-Proliferation Role of the International Atomic Energy Agency: A Critical Assessment." Washington, DC: Resources for the Future.

Schlesinger, James. 1979. Quoted in "The National Energy Act." Washington, DC: Committee of Energy and Natural Resources, U.S. Senate, 96-1, p. 1.

Stanfield, Rochelle L. 1984. "Why Won't the Synfuels Corporation Work? The Real Problem May Be Technology." *National Journal,* June 9.

Tabors, R. D., and P. R. Rogers. 1980. "Energy in Cities." Draft Report to U.S. Department of Housing and Urban Development. Cambridge, MA: Meta Systems, Inc.

# 4 Corporate Blunders

The public sector had no monopoly on dysfunctional decisions in energy system planning. The private sector shared equally in its inability to accept the reality of fundamental changes in the energy economy and to understand the insidious power of faulty assumptions. This chapter examines some of the energy blunders that have occurred within corporate America. The purpose is not to fix blame but rather to understand the assumptions and events that led informed, well-intentioned decision makers to poor decisions.

Some errors were caused by incorrect assumptions, some by resistance to change, some by a lack of understanding of the dynamics of businesses and technology, and still others by a lack of appreciation of the forces of society. The inability to challenge our own assumptions on economy of scale led us to design larger and larger systems, hoping that they would cost proportionally less per unit of capacity. Similarly, we assumed that it was possible to build systems larger and have them perform at least as well as their smaller precursors. Both of these assumptions led to significant strategic errors.

An example of believing that the old way is the only way was American industry's steadfast doctrine that tetraethyl lead in gasoline was an absolute requirement for the kind of performance required to sell automobiles. It was ingenuity that was required, not necessarily lead. In the electrical industry, polychlorinated biphenyl (PCB) was held to be both safe and irreplaceable in our large power transformers and capacitors. Eventually it proved neither safe nor necessary. Again, there was a better solution once the question was asked correctly.

## THE ASSUMPTION OF ECONOMY OF SCALE

Economy of scale is a well-known concept in equipment design and fabrication. For example, in large bypass fan (jet) engines (see Figure 4-1) an increase in size by a factor of two results in a reduction in cost by nearly 30%. In the case of transformers, a similar economy of scale exists. In the case of power electronic equipment, the economy of scale is not quite as pronounced, because part of the increase in rating is accomplished by increasing the number of identical components. Economy of scale for steam turbines is similar to that for power electronics; for example, a requirement for 400 megawatts of low pressure is satisfied with two identical units of 200 megawatts each. Nevertheless, it is not surprising that with such evidence the corporations believed the assumption of economy of scale held for power plant construction also.

John Fisher was the first to point out that there may be no economies of scale in the capital cost of fossil plants (Lee, 1987). In 1975, Fisher analyzed all U.S. coal-fired units coming into service between 1960 and 1972 for which the Federal Power Commission (FPC) had cost data. The FPC data support the view that capital costs per unit of output declined to the mid-1960s, then began to increase sharply. The data show minor influences of adding site preparation costs to the cost of the first unit on a site. But the data show no influence whatsoever of kilowatt rating on cost/kW for units larger than 200 megawatts. (See Figures 4-2 and 4-3.) Adjusting for nonconventional construction, first-unit and second-unit cost increments, and GNP deflator did not significantly alter the conclusions.

Subsequent to Fisher's analysis, an independent study was done by a separate planning group in GE on the same 305 fossil-fired plants analyzed by Fisher, and the basic data source was also the same (Lee, 1987). The data were analyzed in a number of different ways:

$/kilowatt versus size for unadjusted unit size at the national level and the regional level

Adjusted for time, such as GNP deflator with time lag

Adjusted for parameters, such as public versus private, conventional versus nonconventional construction, and coal versus dual fuel fired

The conclusion is obvious: there is a wide spread of capital costs, but no significant economy of scale for units larger than 200 megawatts. (See Figures 4-4 and 4-5.)

The analysis can be applied to nuclear plants. Cost estimates by the Atomic Energy Commission (U.S. Atomic Energy Commission, 1974) based on direct materials and labor indicated a rather strong

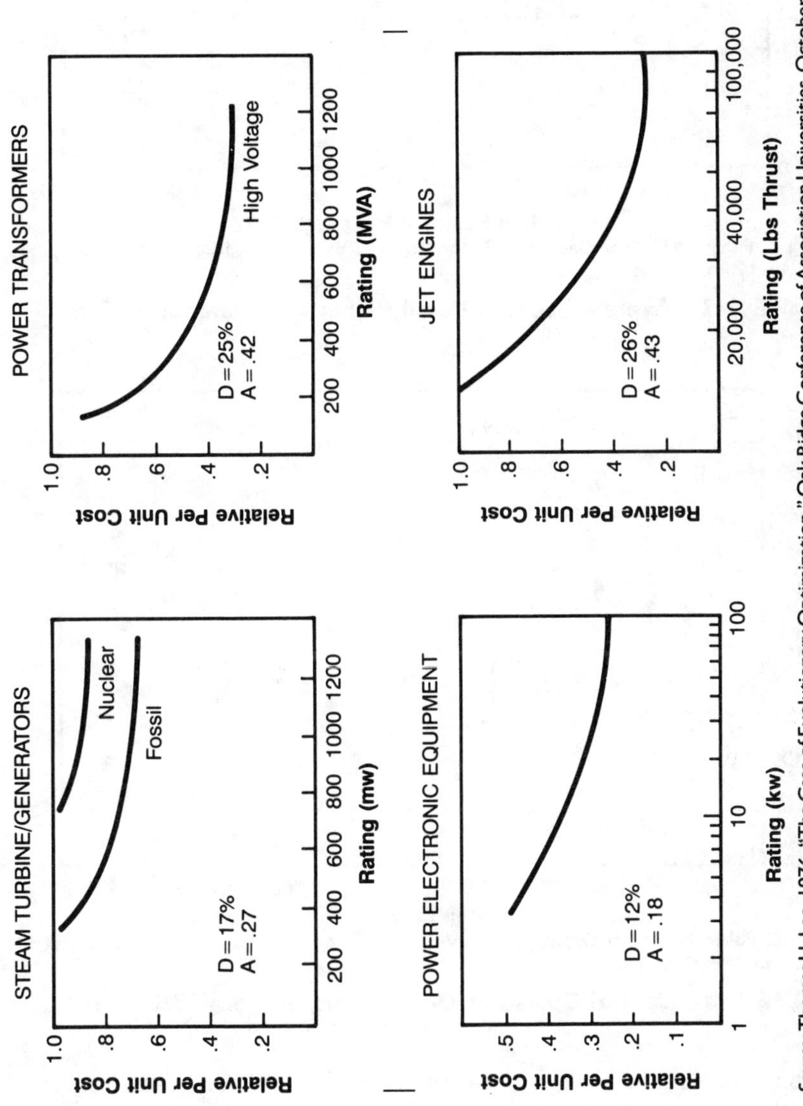


*Source:* Thomas H. Lee. 1976. "The Case of Evolutionary Optimization," Oak Ridge Conference of Association Universities, October 20–21.

**FIGURE 4-1 Economy of Scale for Typical Equipment. See equation 4-2 for definition of A.**

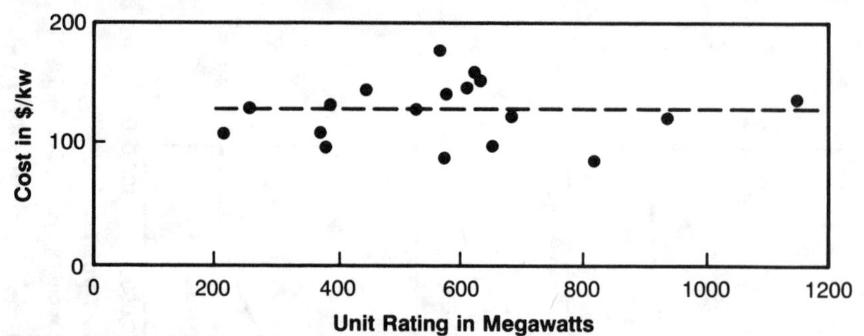


*Source:* J. C. Fisher. 1978. "Size Dependent Performance of Subcritical Fossil Generating Units," EPRI Report, Palo Alto, October.

**FIGURE 4-2 Average Cost of Fossil Plants as a Function of Size.**

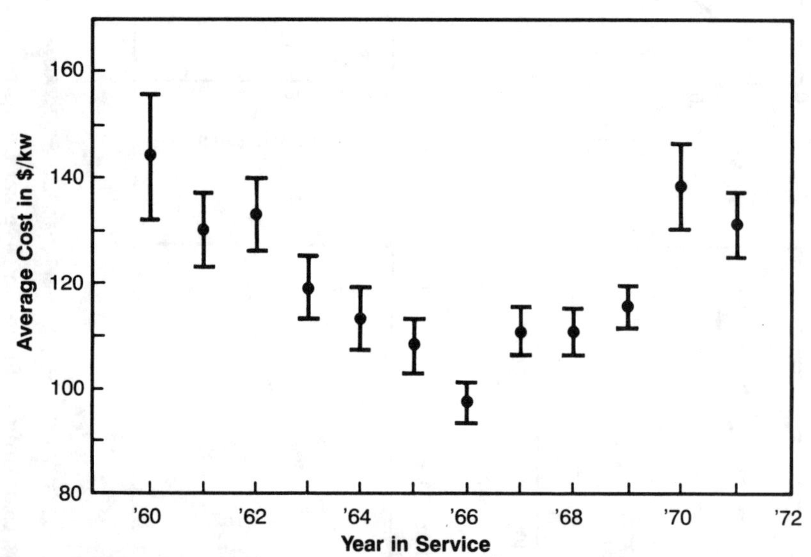


*Source:* J. C. Fisher. 1978. "Size Dependent Performance of Subcritical Fossil Generating Units," EPRI Report, Palo Alto, October.

**FIGURE 4-3 Capital Costs per Unit of Output from 1960 to 1972.**

economy of scale (see Figure 4-6); yet when costs per kilowatt were plotted for plants that went on line between 1972 and 1975, the results showed a different trend (see Figure 4-7). There are many reasons why real economy of scale for nuclear plants is difficult to obtain.

- Plant construction cost is reported as current dollars, uncorrected for inflation.
- Construction costs do not necessarily follow the general pattern of inflation.

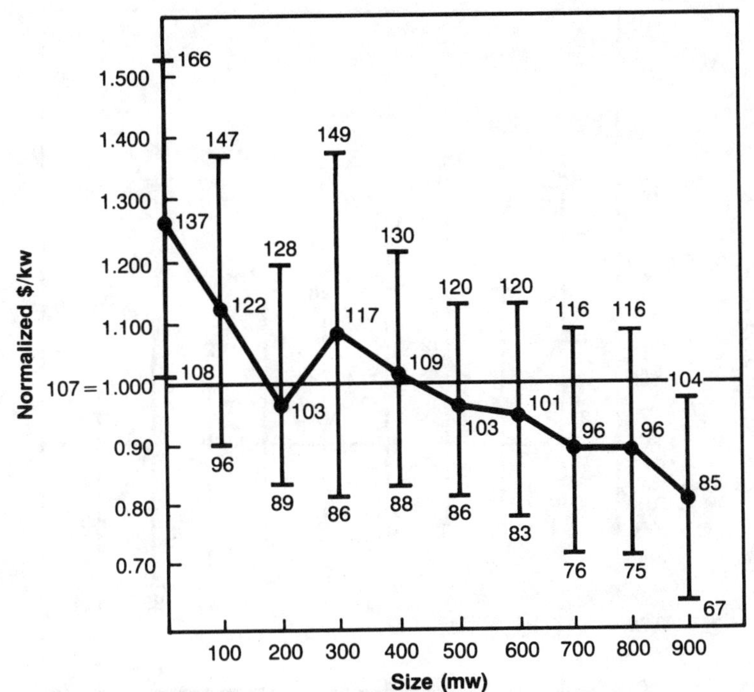


*Source:* T. H. Lee. 1987. "Assessments and Measurements for Energy Conversion Systems," IIASA Working Paper Number WP-87-27, Laxenburg, Austria.

**FIGURE 4-4 All Units—Normalized for Time Parameter and Regional Differences Using GNP Deflator (1980 = 1.0). $/Kw vs. Size. Range Shows ±1 Standard Deviation.**

- The regulatory process was rapidly changing for nuclear plants during the period under consideration. The changes involved back fits, new requirements, and associated construction delays.

To sort out the effects of different factors, a regression analysis was carried out on a data set of thirty-one nuclear plants, the costs for which had been reported in FPC documents. A simple model was developed (see Figure 4-8). In this model, $T_B$ is when construction begins. We assumed that all costs were paid at one time, $T_o$ (half-way between $T_B$ and $T_c$, time for commercialization), and the interest charges were paid between $T_o$ and $T_c$. The simplified equation is then:

$$\text{Cost} = K\left(\frac{R}{1000}\right)^{-A}(1+E)^{(T_o-T_B)}(1+i)^{(T_c-T_o)} \quad \text{Equation 4-1}$$

Where the cost is in $/kilowatt, $K$ is a constant, $R$ is the rating in thousands of kilowatts, $A$ is a measure of the economy of scale, $E$ is the escalation rate for construction cost, and $i$ is the interest rate.

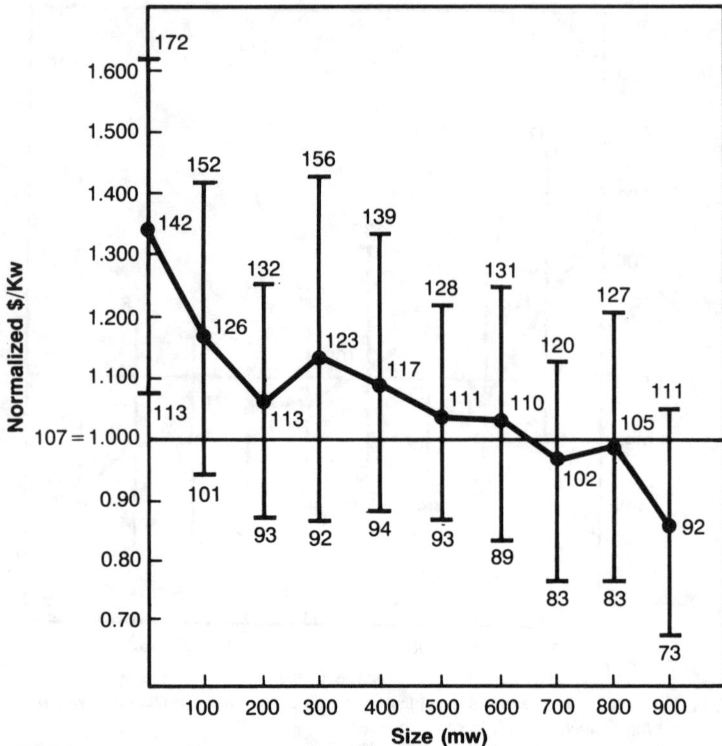


*Source:* T. H. Lee. 1987. "Assessments and Measurements for Energy Conversion Systems," IIASA Working Paper Number WP-87-27, Laxenburg, Austria.

**FIGURE 4-5 All Units—Normalized for Time Parameter and Regional Differences Using Two-Year Lagged GNP Deflator. $/Kw vs. Size. Range Shows ±1 Standard Deviation.**

The interest rates used were close to the actual level during the period. The values derived from the analysis are:

$$K = 212$$

$$A = 0.49$$

$$E = 0.36$$

Several conclusions are immediately obvious:

- If one considers only the cost of materials and direct labor, and neglects the effects of escalation and interest rates, nuclear plants do exhibit rather strong economies of scale in direct material and labor costs. Of course, that was the basis of the estimates by AEC and the rest of the industry.

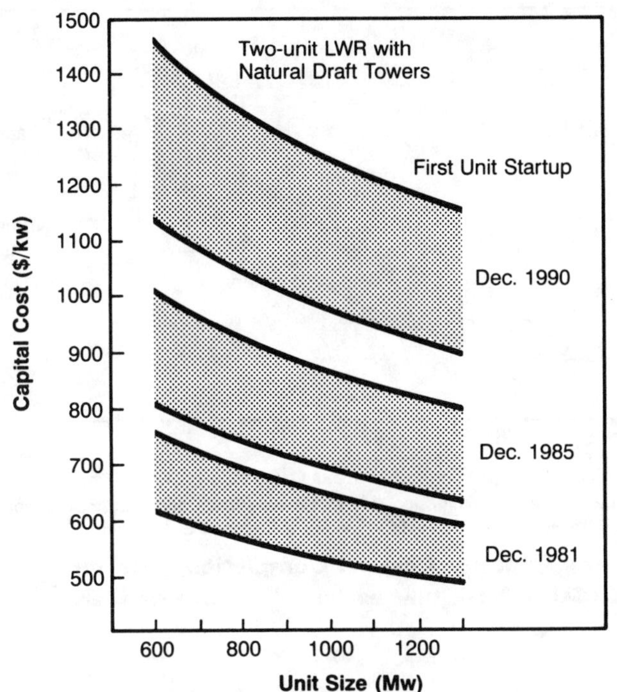


*Source:* U.S. Atomic Energy Commission. 1984. "Power Plant Capital Costs, Current Trends and Sensitivity to Economic Parameters," WASH-1345, Washington, DC, October.

**FIGURE 4-6 Range of Projected Costs for LWR Plants.**

- The escalation rate per year in construction costs was alarmingly high. Looking back at that period, we can identify many causes: the *Calvert Cliff* case,[1] the emphasis on emergency core cooling system, and so forth. Given the escalation that has been experienced, nuclear plants are now economically unattractive.
- Fisher noted that construction time may also be a function of size. This could introduce a *diseconomy* of scale.

The studies on both nuclear and fossil plants cast serious doubts on the validity of the economy of scale assumption. Bigger facilities are not necessarily less expensive per unit of capacity. In fact, they may be more expensive. The problem was that decision makers believed that the larger plants would be inherently less costly. This was the basis of their decisions favoring larger facilities. The result was that many larger systems actually cost more, not less. To be fair, the data on costs of the larger plants were available only after the decision to build them and the actual construction had taken place. Decision makers could not have had this information.

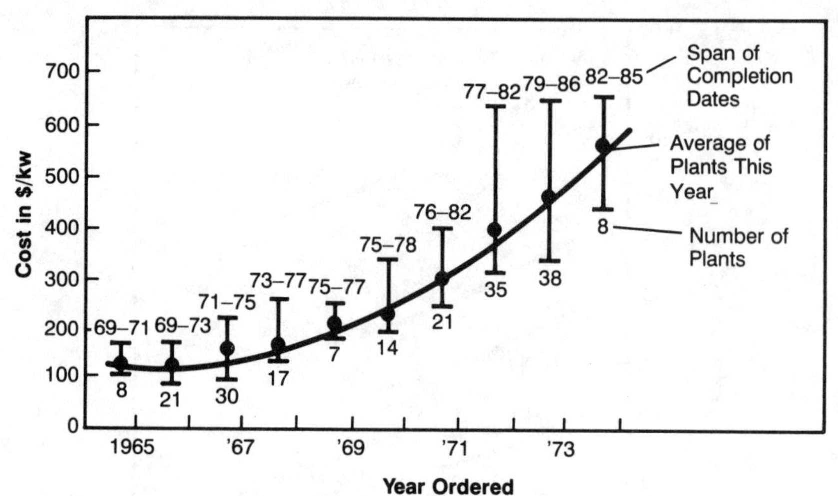


*Source:* U.S. Atomic Energy Commission. 1974. "Power Plant Capital Costs, Current Trends and Sensitivity to Economic Parameters," AEC-1345, Washington, DC, October.

**FIGURE 4-7 Capacity Cost vs. Completion Date for U.S. Nuclear Plants (1960 $US). Note that earlier plants are smaller.**

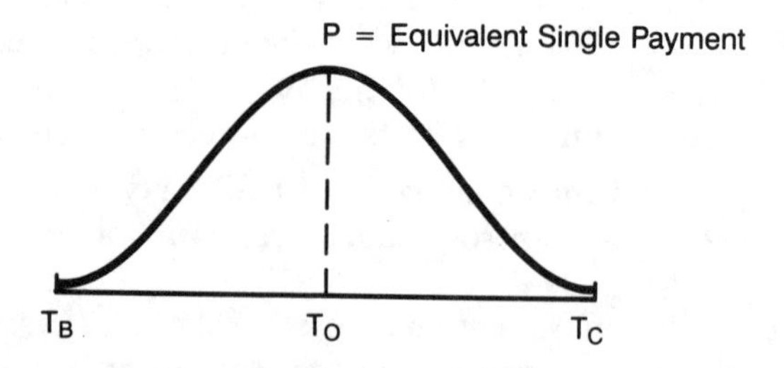


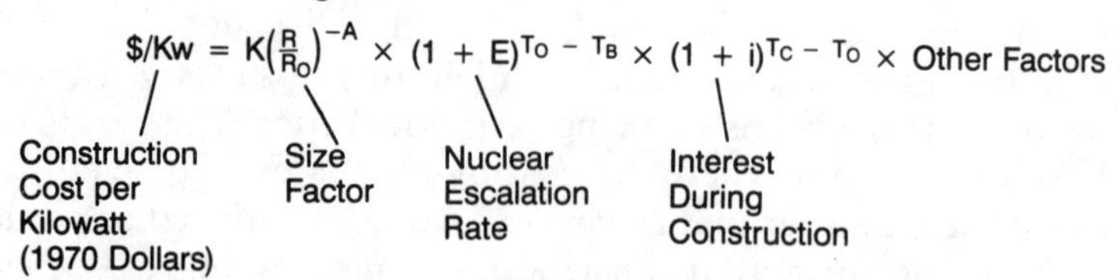


**FIGURE 4-8 Cost Factors of Nuclear Power Plants.**

## THE ASSUMPTION OF CONSTANT CAPACITY FACTOR

A second assumption relating to performance and size in large investments in the electric power sector was that of constancy in plant capacity factor.[2] In plain English, this means that power plants of all sizes were assumed to operate, over the long term, at the same average percentage of their maximum capacity, independent of size. In practically all generation-planning studies, the capacity factor is assumed not to be size dependent. The most popular number was 0.7 regardless of size. Fisher pointed out that the assumption may be wrong. (See Figure 4-9.)

There are a number of reasons why capacity factor is, in fact, a function of size.

- Large plants may have more identical components, each with an independent probability of failure. Therefore, the total system has a higher probability of failure than any one component.
- Large plants require a longer time for each maintenance.
- Random probability of failure is a function of operating stress. Larger equipment may operate at higher stress levels.[3]

In a 1976 study, nuclear plant capacity factor was evaluated for operating plants (Lee, 1987). (See Figure 4-10.) No conclusion can be drawn from such a plot because of the way capacity factors are

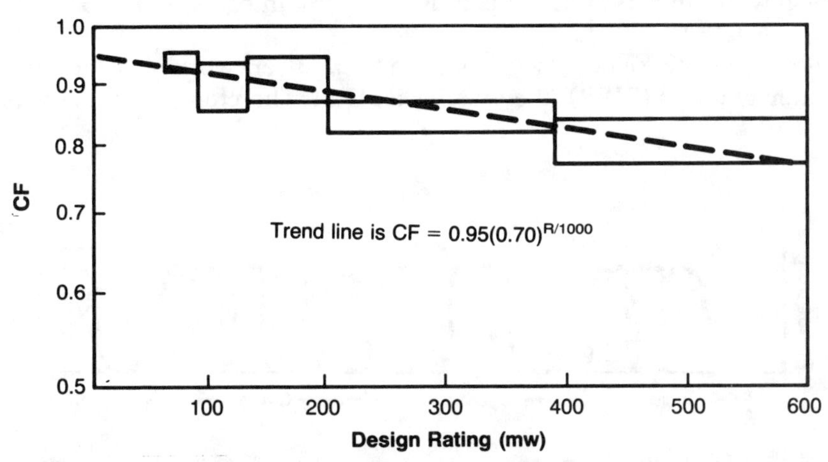


The horizontal bars represent the range of ratings in a group of units.
The vertical position of the bars represents capacity factors for the units, the average based on rates from third through sixth year of operation.

**FIGURE 4-9 Capacity Factor of Fossil Steam Plants.**

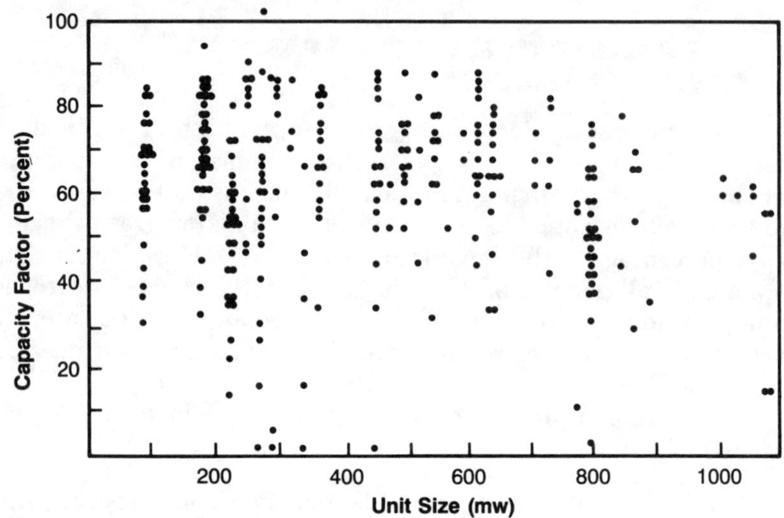


*Source:* T. H. Lee. 1987. "Assessments and Measurements for Energy Conversion Systems," IIASA Working Paper Number WP-87-27, Laxenburg, Austria.

**FIGURE 4-10 Capacity Factor for Operating Nuclear Plants.**

reported. If one computes the capacity factor for completed cycles (see Figure 4-11), as suggested by Fisher, the trends become clearer. Fisher also omitted the first cycle, and only used units between 400 and 800 megawatts in size. For his data he relied on the Nuclear Regulatory Commission gray book.[4] For an increase of 1,000 megawatts in rating, there is a drop of 40% in capacity factor. (See Figure 4-12.)

When these results were shown to engineers working on boiling-water reactors (BWR), they were skeptical. Therefore, an effort was

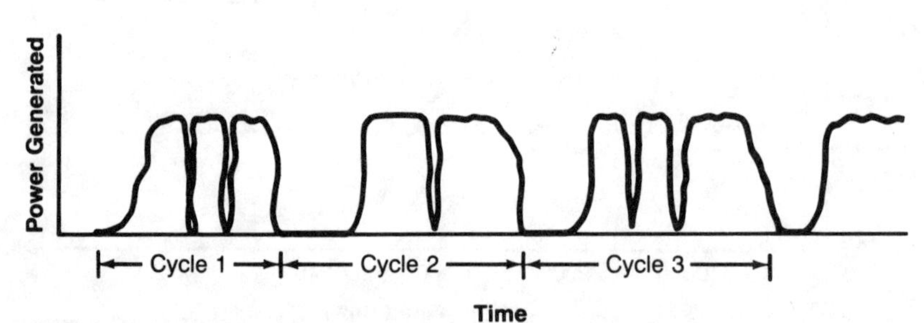


*Source:* T. H. Lee. 1987. "Assessments and Measurements for Energy Conversion Systems," IIASA Working Paper Number WP-87-27, Laxenburg, Austria.

**FIGURE 4-11 Capacity Factor for Completed Cycles in Nuclear Plants.**

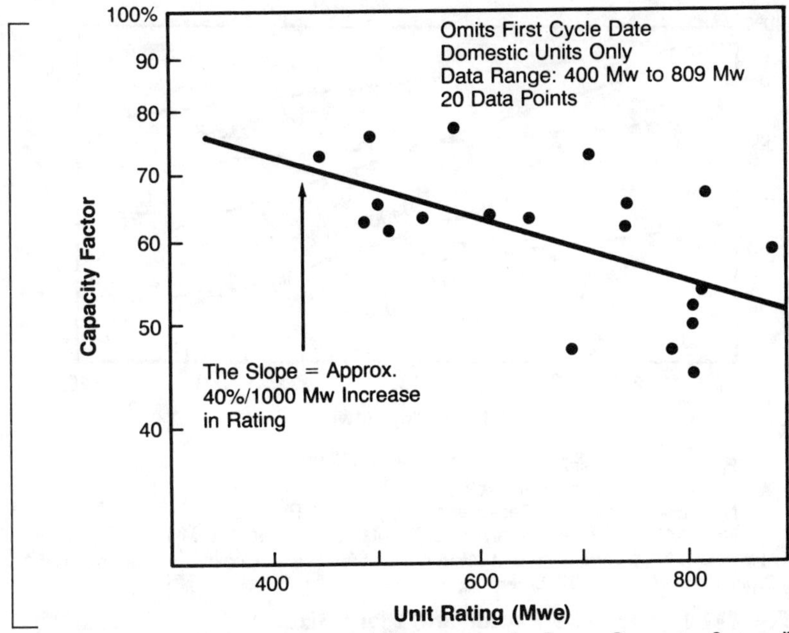


*Source:* T. H. Lee. 1987. "Assessments and Measurements for Energy Conversion Systems," IIASA Working Paper Number WP-87-27, Laxenburg, Austria.

**FIGURE 4-12 Capacity Factor of Nuclear Plants with High Ratings. Analysis by J. C. Fisher.**

made by the engineers to study in detail all causes for the loss in capacity factor for all BWRs on-line in 1974–1975. (See Table 4-1.)

**TABLE 4-1 Contributors to BWR Plant Capacity Factor Reductions, 1974–1975**

| *Out of Service* | *~ Size* | *Not ~ Size* | *Total* |
|---|---|---|---|
| Nuclear Steam Supply System | 8.3 | 14.2 | 22.5 |
| Balance of Plant | 1.5 | 1.5 | 3.0 |
| Brown's Ferry Accident | 0 | 6.6 | 6.6 |
| Regulatory Intervention | 0 | 3.0 | 3.0 |
| Other | 0.25 | 0.25 | 0.5 |
| Derate | | | |
| Power limit due to fuel | 0.7 | 8.3 | 9.0 |
| Power reduction (utility option) | 0 | 1.5 | 1.5 |
| Total | 10.75 | 35.35 | 46.1 |
| Resulting Capacity Factor | | | 53.9 |

*Source:* General Electric. 1976. Unpublished analyses carried out for Thomas H. Lee, San Jose, CA.

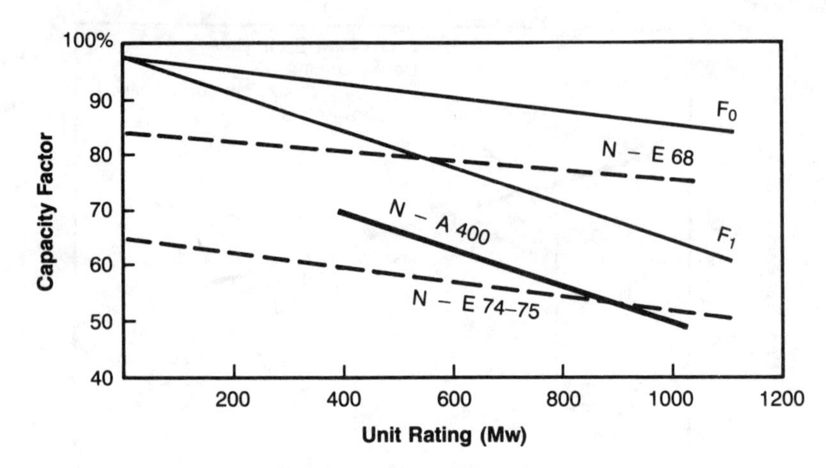


| | |
|---|---|
| $F_0$ | Fossil Plant with Perfect Boiler |
| $F_1$ | Fossil Plant Actual |
| N – E | Nuclear Designer's Analysis and Prediction |
| N – A400 | Fisher's Analysis for Plants Larger than 400 Mw |

*Source:* T. H. Lee. 1987. "Assessments and Measurements for Energy Conversion Systems," IIASA Working Paper Number WP-87-27, Laxenburg, Austria.

**FIGURE 4-13 Capacity Factor and Plant Size.**

The total loss in capacity factor was 46.1%. Size effects can account for 10.75%. That is, the engineers can only attribute 10.75% of the loss in capacity factor to large size. But design engineers for electromechanical equipment usually do not consider the dependence of random materials failure on operating stresses in scaled-up units (although electronic engineers know this well). Taking that into consideration, the planners concluded that the size effect may be responsible for 15% to 20% of the loss in capacity factor for an increase of 1,000 megawatts in rating. The study suggests that the effect of size on capacity factor can fall within a rather wide range. (See Figure 4-13.)

## THE IMPLICATIONS OF TWO WRONG ASSUMPTIONS

The traditional equation for economy of scale is:

$$\$/\text{kilowatt} = \text{constant} \times R^{-A} \qquad \text{Equation 4-2}$$

where $A$ is a measure of the economy of scale and $R$ is rated capacity. If one assumes a linear decline in capacity factor as a function of size, then the effective capital cost,

$$\frac{\$/\text{kilowatt}}{\text{Capacity Factor}}$$

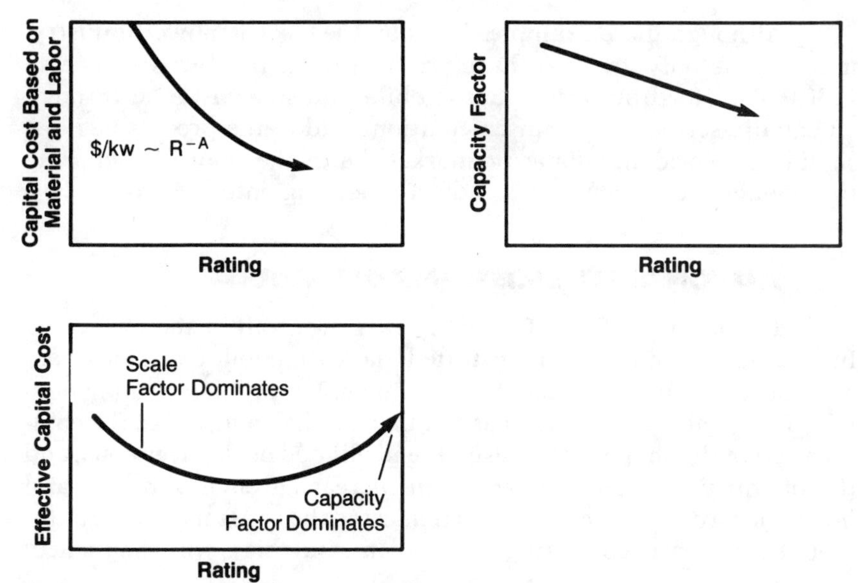


**FIGURE 4-14 Cost, Size, and Capacity Factor.**

will have a size dependence (see Figure 4-14). This effect was never considered in generation planning. We will return to a discussion of further implications in the next chapter. Suffice it to say that the industry's strategy was certainly not what Figure 4-14 would teach us to do: search for the optimum. Instead, it pushes economy of scale to the limit, increasing size over time at an average rate of 60 megawatts per year, as mentioned before. It should be noted that this fault was industrywide. Manufacturers, architect–engineering firms, utilities, public utility commissions (PUCs), AEC, Nuclear Regulatory Commission (NRC)—no one questioned the assumptions on which the industry proceeded.

The results of adhering to the two assumptions concerning plant operations for too long is now fairly clear in the electric power system in the United States and much of Western Europe. The units most recently brought into service are larger than earlier units of the same technology, their effective capital costs per unit of capacity and per unit of output are appreciably higher, and their operating performance is no better and often poorer than the smaller units, in spite of the recent technical efforts to improve the performance of nuclear plants. No one has accurately calculated the financial implications of the unanticipated results, but the financial pains suffered by the electric utility industry bear witness to how severe they must be.

Although these examples focus on the electric power industry, it is not the only group to have erred through the 1970s and early 1980s. The North American automobile industry has gone through an equally serious set of miscalculations and in the process has lost significant world and domestic market share. The error was ignoring the signals of change while steadfastly holding onto the past.

## AUTOMOBILE EMISSION CONTROL

In the early 1950s, Professor Haagen-Smith of the California Institute of Technology demonstrated that automobile emissions were a major contributor to photochemical smog. The problem was particularly important in areas such as Southern California, where the combination of the shape of the basin, the likelihood of air inversions, and the quantity of emissions led to an increasing environmental and health hazard. California officials requested the assistance of the automobile industry in correcting the problem, and this is what happened.

- The industry first reacted by denying it played any major role in the emission problem.
- In 1955, a cross-licensing agreement granted manufacturers royalty-free access to any emission control patents owned by firms within the agreement. The agreement reduced the incentive to pursue research in emissions because it removed any potential competitive gain for the developer.
- In 1963, California passed legislation to provide direct incentives, intended to motivate technological development of low-emission automobiles.
- In June 1964, emission-control devices developed by four auto parts manufacturers who were outside the automobile industry were certified by the state of California.
- In August 1964, automobile companies announced that exhaust-control devices would be provided in two years.
- Standards set in a 1970 amendment to the Clean Air Act mandated a 90% reduction in emissions of the oxides of nitrogen ($NO_x$) by 1976.
- Protests by industry that these standards were impossible to meet and might irrevocably cripple the automobile industry forced the Environmental Protection Agency (EPA) to grant a one-year delay in 1973.
- At roughly the same time, the Honda CVCC engine was certified as meeting the 1976 standard!

The attitude that it is better to ignore the problem than to face it was clearly carried forward by U.S. automobile manufacturers. But the problem did not go away, and the demand for improvement remained. A remedy *could* be found. While domestic manufacturers stonewalled, Honda did it. The industry finally was forced to face the fact that emission improvement was possible, and that foreign competition could make it happen—within the United States.

## FLUE GAS SCRUBBERS

The next example is as clear as the first. Air quality deterioration within metropolitan regions was seen to be a product of—among other things—large-point sources, specifically power plants and industrial boilers. The process that culminated in the Clean Air Act included extensive testimony from the electric utility industry. The actions and arguments surrounding the electric power industry's opposition to reducing emissions from generation plants looked remarkably like the automobile industry's opposition to the reduction of automobile emissions. Let's review the sequence of events.

- In 1970, Congress directed the EPA to set new source performance standards (NSPS) on emissions from new, large stationary sources such as power plants.
- In December 1971, EPA promulgated its NSPS for coal-burning power plants and identified scrubbers as an "adequately demonstrated" technology. This step was critical because the criteria were based on available technology.
- A number of midwestern utilities switched to low-sulfur western coal with a resulting loss of work by coal miners in Appalachian and midwestern coal fields.
- During 1975–1977, an unusual alliance developed between environmentalists and high-sulfur (eastern) coal interests. For totally different reasons, both pushed for legislation that would require a percentage reduction in the oxides of sulfur ($SO_x$) instead of emission limits.
- In February 1978, Georgia Power Company officials wrote President Carter and Congress opposing the concept of required scrubbing on all power plants (Haskell, 1982).
- In June 1978, the Business Roundtable met with President Carter to oppose full scrubbing.
- Final NSPS regulations were published in 1979.
- Between June 1979 and 1981, a series of legal battles took place, with all sides fighting the EPA.

- In April 1981, a U.S. court of appeals affirmed EPA's authority to set NSPS.

As was the case with the auto emission control systems, Japanese competition entered before the case was closed. In January 1979, Senator Jackson (D, Washington) called for the convening of a task force to study the Japanese scrubber technology. In 1979, the Electric Power Research Institute (EPRI) began a study on second-generation Japanese scrubber technology and U.S. manufacturers started serious discussions with Japanese suppliers on licensing their technologies. In addition to advanced technologies on scrubbers, Japan also developed new catalytic converters for gas turbines and U.S. suppliers used them to remove oxides of nitrogen from the flue gas (so-called $deNO_x$ systems).

What was the U.S. industry doing during this period? It was digging in its heels against the irresistible forces of change. The air quality debate in the United States focused primarily on why the use of scrubbers was not a viable possibility, why it would cost too much, and why it would not work effectively. Nevertheless, the problem did not go away, and a consensus finally has emerged. The air problem can be solved. The point is not only that the United States has lost the lead in air quality control technology but also that an opportunity for progress—for moving toward what were being defined as socially viable goals—was lost, for lack of vision.

Two other cases illustrate the resistance to change based on social needs. Although the cases did not have the effect of undermining the U.S. competitive position, they did significantly and justifiably shake even further the confidence of the U.S. population in the judgment, credibility, and social conscience of industry.

## TETRAETHYL LEAD

In the mid-1960s, it became clear that a major source of lead pollution in both air and water, as well as residual pollution on the land, was being caused by the tetraethyl lead additive in motor gasoline. The purpose of the additive was to eliminate engine knock. Octane rating is a measure of gasoline's ability to eliminate knocks. The lowest-cost means of increasing octane rating includes the addition of tetraethyl lead.[5]

At the time all motor gasoline contained lead. The questions asked were, Could you make a fuel that would allow you to operate a vehicle satisfactorily without lead? Was deterioration in performance worth the cost? The oil companies knew the answers. They said it over and over at high volume. You could not produce a fuel that would burn satisfactorily in a gasoline engine without lead additives. Du Pont and the Ethyl Corporation, the leading producers of tetraethyl lead, concurred. From their perspective, a reduction in lead meant a significant increase in the manufacturing costs of gasoline. The fact is, of course, that at least part of the improvement in octane rating could be achieved by the use of higher-cost refining technology, such as the addition of new refinery processing equipment. Further, there was the cost of new tanks and new pumps in every station, to handle a new grade of unleaded gasoline.

The automobile manufacturers resisted because the performance of larger automobiles would be affected dramatically.

What were the results? As usual, shouting did not make the problem go away; it only wasted time in reaching the inevitable solution: lead-free gasoline and redesigned engines; new equipment in the refineries to provide the octane-improving processing; a slightly lower overall refinery product yield; a slight reduction in octane rating levels and engine performance; and a slight increase in the cost of gasoline, which was largely passed on to the consumer. These changes resulted in dramatically reduced lead-based pollutants in the air and in surface and ground water. It is interesting to note that the lead from gasoline was never conclusively shown to have any effect on individuals chronically exposed to it in this form, such as traffic police and tunnel guards. Society was convinced, however, that lead oxides were detrimental. Taking them out of gasoline was the intelligent action, not because of the health of those chronically exposed to the fumes, but rather because of the damage that results to the environment in general by the lead oxides buildup.

The U.S. response was dramatically different from that of the countries in Western Europe. In the United States, the social goal was to change from leaded to unleaded fuels at an acknowledged cost in price, yields, and performance. The cost to the ultimate consumer was masked by the fact that gasoline mileage was improving with lighter new automobiles during the changeover period. Society got what it wanted *despite the fact that expert opinion argued that it could not be done.* Only recently has Europe begun to take effective action in this direction. It should not have to repeat the tumult that accompanied the change in the United States.

## POLYCHLORINATED BIPHENYL (PCB)

PCBs are liquid compounds with attractive heat transfer and electric properties. As an electrical insulating liquid, they have high dielectric strength and are not flammable in the presence of an electric discharge. Therefore, they were very popular for use in transformers and capacitors. Because of their heat transfer characteristics, PCBs were also extensively used in heat exchangers.

In the late sixties, concern began to mount on the probable environmental impact of PCBs. They are nonbiodegradable and adsorb to particles of soil or mud on the ground or the bottom of rivers and estuaries into which they find their way. Both manufacturers and users of PCBs claimed that they were not harmful to human health. They could cause minor skin rash and eye irritations, but that was all. The response of the electric power industry was that PCBs were here to stay because their value far outstripped their danger to society and any danger to society could be handled through isolation.

The situation changed in the early 1970s when leaks in a heat exchanger caused massive food poisoning in Japan. Shortly thereafter, PCBs were completely banned in Japan. The wave of public opinion against PCBs quickly spread to the United States. Massive campaigns were launched to remove PCBs from all uses. Areas that had received significant polution from PCBs, such as the Housatonic River (Pittsfield, Massachusetts) and the harbor at New Bedford, Massachusetts, became targets for major litigation demanding industrial cleanup. The results of the focus on PCBs in the United States is interesting when viewed now, nearly twenty years later.

- PCBs are no longer used in either capacitors or transformers.
- Methods of high-temperature incineration have been developed to destroy PCBs.
- Industry found substitutes for PCBs in capacitor applications that have proven superior in both electrical and thermal settings. As a result, the sizing and quality of capacitors on power systems have improved.

In the end, the industry was forced to respond with innovative solutions to remove and replace PCBs in both the environment and operating equipment. Today, even though there still is no convincing proof that PCBs are carcinogenic, public concern brought about the change at a high cost to the industry. Once again, part of the high cost was not U.S. dollars and cents but in a much different currency: the loss of public confidence in an industry found to be resistant to change and not socially responsible.

## MISMANAGED TECHNOLOGY

Technological development is a process to which we will return in a later chapter. Understanding the dynamic of technologies is critical to making decisions, either to proceed with one technology or to slow progress on another. During the 1970s, the management of technology development became a muddle of private- and public-sector activity. A number of errors were caused by the lack of understanding of this important element. The two examples that follow are taken from the private sector.

### A Case for Technology Assessment: Magnetohydrodynamics (MHD) Power Generation

Traditionally the costs of electricity produced—busbar[6] costs—were the most important consideration in assessing the value of alternate electric conversion systems.[7] The costs are calculated from the levelized cost[8] of capital and the expected cost of fuel plus operating and maintenance costs. For well-developed and well-understood technologies, the busbar cost numeraire can be very useful. Even here, we should point out that the operation and maintenance costs of some plants (nuclear, for example) are often grossly underestimated.

For new technologies, there are many issues beyond the knotty questions of the cost of electricity and initial capital costs. Two of the most important issues are technical feasibility and cost and time required for research and development. Other considerations, such as safety, ease of maintenance, and reliability are important but cannot be handled quantitatively by economic calculations.

The need for better ways to assess energy conversion technologies was felt especially strongly after the 1973 oil embargo. NASA, EPRI, and the Energy Research and Development Administration (ERDA) sponsored separate projects to compare different energy conversion systems (Lee, 1987). In one study, three evaluation techniques were proposed: levelized cost of electricity, net present worth, and direct weighting. Let us review the methods quickly as background to the MHD case.

To measure the cost of electricity using the levelized cost method, estimates of capital cost and operating costs are needed. In order to reduce uncertainties, the study team relied on "experts" (i.e., the promoters) to estimate the cost of that part of the plant which required expert opinion. Architect and engineering firms experienced in power plant design and construction were used to estimate the costs associated with the "balance of plant." The purpose was to ensure consistency. This method also made an assumption on the reliability of the plant, for example, its capacity factor. The assumption probably was the most important uncertainty in the analysis of the system, but its viability was never questioned. Because of the uncertainties, the comparative costs have a wide margin for error; differences less than 25% should be considered in the noise. This was not, however, the general attitude. Advocates of a given technology pushed their pet idea by arguing that "15% savings in electricity costs could be achieved with . . ." For a technology that has never seen the market, such differences are without significance.

The second method proposed for the evaluation was the present worth of the technologies. The evaluation is again difficult because of the assumptions required: for example, predictions of time and cost needed for research and development, market penetration in the face of both existing and new technologies, and the potential savings from the introduction of the new technology. Thus the conclusions from the calculations—so heavily laden with guesstimates and assumptions—must also be taken with a grain of salt.

The third method, direct weighting, was designed to give consideration to all other variables that are intangible in the busbar cost and present worth calculation. Altogether twenty-six different considerations were proposed. The most significant of these were:

- cost of research and development
- time required for research and development
- probability of development success
- hardware and materials availability
- industrial capability for manufacture
- environmental degradation

A rather elaborate method was devised to calculate the statistical weights of the variables (Lee, 1987). The weighting was done by composing a list of corporate objectives and then assigning a relative importance to each of the intangible variables with respect to the objectives. Then, a rating was given to every conversion system for every intangible variable. The product of the rating and the statistical weight for that variable then yields a merit rating for the specific system in relation to the variable. The sum of the ratings gives a figure of merit for that conversion cycle. All of the cycles studied were then displayed on a two-dimensional matrix (see Figure 4-15).

The technologies are spread over the matrix. Technologies that were well developed, such as gas turbines, had a high cost of electricity (COE) (bad) and high figures of merit (good); technologies that we knew little about such as open cycle magnetohydrodynamics (MHD)

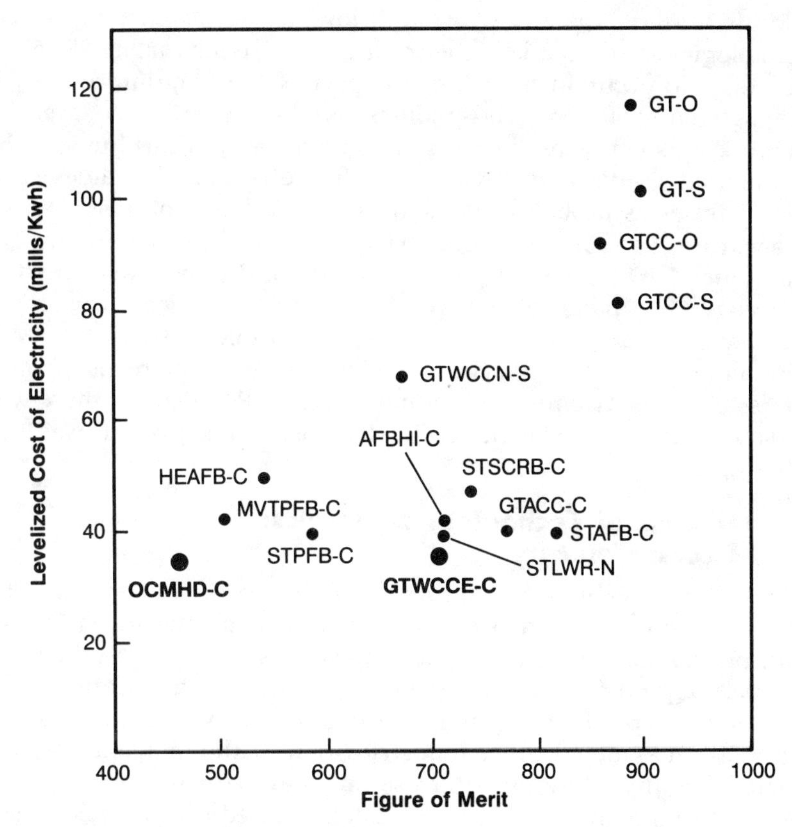


**Plant Type**:

| | | |
|---|---|---|
| STLWR | - | Steam Light Water Reactor |
| STSCRB | - | Steam Conventional Furnace Scrubber |
| GTCC | - | Conventional Gas Turbine Combined Cycle |
| GT | - | Conventional Gas Turbine |
| STAFB | - | Steam Atmospheric Fluidized Bed |
| GTWCCE | - | Advanced Gas Turbine (Water Cooled) Combined Cycle LBtu Gasifier |
| OCMHD | - | Open Cycle MHD |
| STPFB | - | Steam Pressure Fluidized Bed |
| HEAFB | - | Closed Cycle Helium Gas Turbine Organic Bottoming AFB |
| MVTPFB | - | Metal Vapor Topping Cycle PFB |
| GTWCCN | - | Advanced Gas Turbine (Water Cooled) Combined Cycle Liquid Fuel |
| GTACC | - | Advanced Gas Turbine (Air Cooled) Combined Cycle LBtu Gasifier |
| AFBHI | - | Advanced Steam (1200 F Reheat), AFB |

**Fuel Type:**

| | |
|---|---|
| N - Uranium | O - Oil |
| C - Coal | S - Semiclean Liquid Derived From Coal |

*Source:* T. H. Lee. 1987. "Assessments and Measurements for Energy Conversion Systems," IIASA Working Paper Number WP-87-27, Laxenburg, Austria.

**FIGURE 4-15 Cost of Electricity (Base Load) and Figure of Merit.**

exhibited an attractive COE and low figure of merit. Clearly technologies that would be the most attractive for further development should occur in the southeast part of the diagram: low COE and high figure of merit. Thus industry might be expected to intensify its work on such technologies as the advanced gas turbine and the pressurized fluidized bed. Since the figure of merit includes such uncertainties as probability of success, a low figure of merit should be a warning flag. One should seriously question the wisdom of investing in the MHD system for power generation. But that was not what happened. The industry charged ahead with the project. Now, more than twenty years later, there is still no operable MHD system. Millions and millions of dollars have been wasted. Even more disappointing is that after twenty years of unsuccessful R&D effort, there was still a strong push for a commercial demonstration as recently as four years ago.

## A Case of Technology Dynamics: Excess Capacity

When a technology approaches maturity, the industry usually faces the serious problem of excess capacity. Today, the industrialized world is plagued by excess capacity in steel, ships, textiles, the manufacture of power conversion equipment, electrical apparatus, and a host of other products. In the next chapter we will show how the same fate may befall other industries that at the moment are considered healthy. Here, we will use an example from the electrical industry to illustrate the type of mistake caused by a lack of understanding of the dynamics of technology.

In the first half of 1974, the electric utility industry in the United States ordered 140 gigawatts[9] of generation additions. This was equivalent to about 25% of the then-total generating capacity in the country. Even though the electric demand had been growing steadily at 7% per year for several decades, such an order rate was still much too high for a given year. People in the industry knew that the surge of orders was partly a protective action against potential price increases and that a 25% increase would be needed in the future. To major suppliers like GE, the increase was a real indication of their customers' faith in the future. Proposals were made to expand its generator-manufacturing facility, a $300 million project.

Fortunately, that expansion never took place because the following question was asked: Where did the demand growth come from in the past, and where will it come from in the future? Fisher (1974) pointed out that for well-established end uses such as motor drives, the growth in electricity demand is about the same as the GNP. It was the new end uses such as television, computers, and air condi-

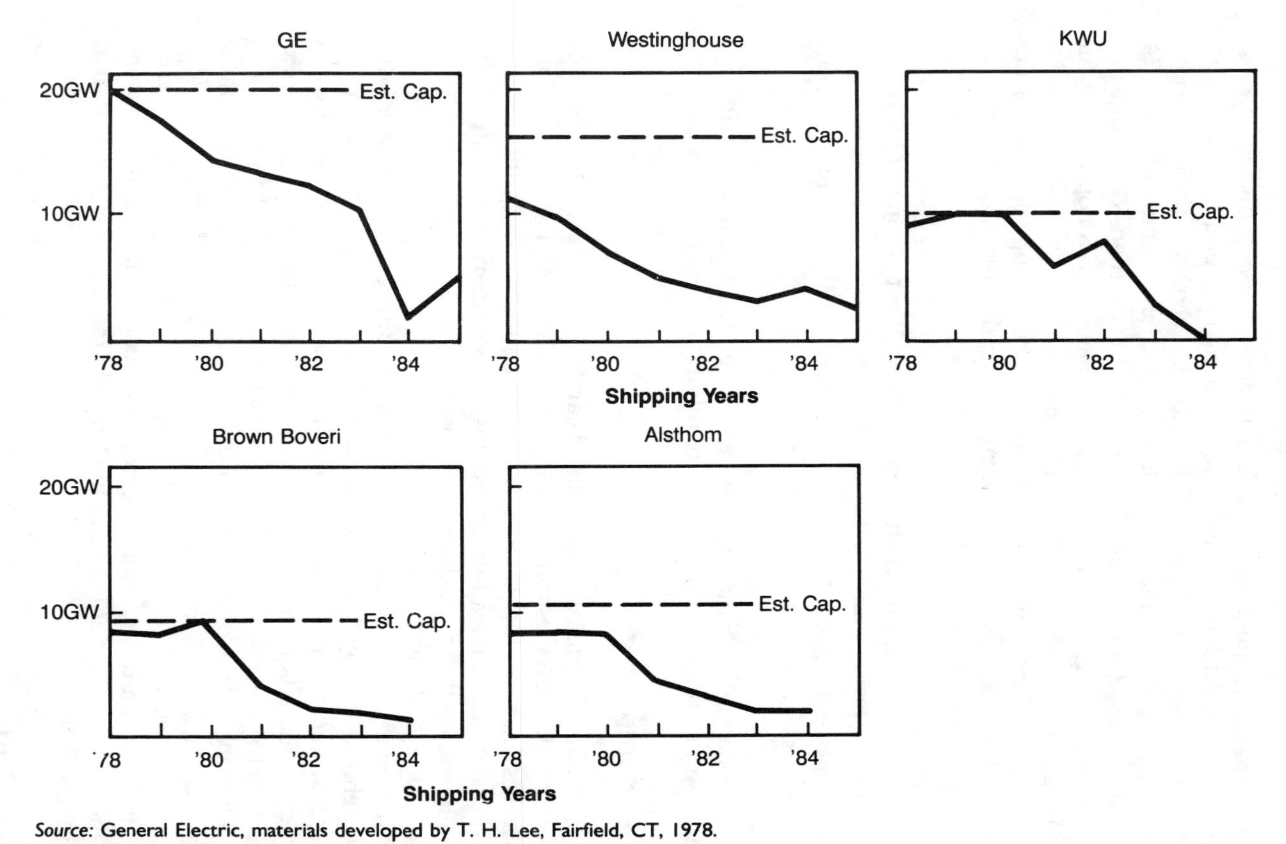


*Source:* General Electric, materials developed by T. H. Lee, Fairfield, CT, 1978.

**FIGURE 4-16 Steam Turbine Manufacturers' Backlog in 1978.**

tioners that sustained the growth between the 1950s and the 1970s. But in the mid-1970s, there were no such new end uses in sight. Because of Fisher's analysis, GE did not proceed with the expansion. The postponements and cancellations in the second half of 1974 proved the wisdom of that decision. But, even then, GE could not prevent the excess capacity because expansion was a popular step in the industry even though it was in the mature stage of the life cycle. For example, from 1978 to 1984, major turbine suppliers had a decreasing backlog, and below-capacity utilization for a few years. (See Figure 4-16.) And, yet:

- In the 1970s, the Kraftwerk Union in West Germany built one of the largest turbine factories in the world.
- Westinghouse built new turbine facilities in the south at about the same time.

Today, all the major manufacturing facilities are starving for work. Some are closed and some are relying on orders for spare parts.

## SUMMARY

In this chapter we have looked carefully at some of the mistakes made by the private sector during the 1970s and 1980s. Some are mistakes in planning because of false assumptions. Some are the result of inadequacies in methodologies. Some are the result of the deeply rooted desire to defend the status quo.

One must not underestimate the forces of an extant system. Tolstoy said, "I know that most men, including those at ease with problems of the greatest complexity, can seldom accept even the simplest and most obvious truth if it be such as would oblige them to admit the falsity of conclusions which they have delighted in explaining to colleagues, which they have proudly taught to others, and which they have woven—thread by thread—into the fabric of their lives" (Gleick, 1987, p. 38).

The causes of these blunders are, in many ways, generic; they are not limited to the energy sector. Some of them contribute to one of the most important issues in the United States: loss of international competitiveness.

In the next chapter we will look at what went wrong with our energy business.

## NOTES

1. In the *Calvert Cliff* case in 1971, a federal court of appeals held that the Atomic Energy Commission must consider environmental issues in licensing proceedings, even when no party or intervenor raises such issues. Before then, the AEC had no power to deal with environmental issues.

2. Capacity factor is (energy actually delivered, kWh in a year)/(maximum output/hour × 8,760 hours in a year).

3. As an example, a one-watt resistor operating at one-half watt will likely outlast a one-watt resistor operating at one watt, but will likely fail earlier than a ten-watt resistor operating at five watts.

4. A publicly available document that contains the operating performance of all U.S. nuclear plants.

5. There are two ways of dealing with the removal of tetraethyl lead from gasoline. The first is to lower the octane rating of gasoline, by redesigning automobile engines with lower compression ratios, and a concomitant reduction in both performance and mileage. The second is to increase the octane rating of the gasoline. This can be accomplished by using more costly refining methods, as well as more expensive and less effective additives. Actually, we are currently pursuing these directions simultaneously. The effect on the consumer is higher cost for gasoline of a given quality, and higher cost of an automobile of given performance.

6. That is, cost delivered to the boundary of the power plant.

7. Energy conversion systems are processes that convert the physical energy in a substance (a lump of coal) into a form that can be better used (electricity).

8. The cost of electricity has three components: the costs of capital, fuel, and operation and maintenance. The costs vary from year to year for a number of reasons. Depreciation methods will alter the cost of capital; inflation will alter the cost of operation and maintenance; and fuel cost is subject to variance. In order to compare the economics of different generating systems over their respective lifetimes, comparable values are needed. The levelized cost approach provides such values. For example, if we have a cost for every year of the twenty-year life of a unit, we can compute the present value of those twenty numbers. Using the same discount rate, we can then calculate the twenty equal payments that would result in the same present value. The amount of each of the equal payments is the "levelized cost." Although this method has the advantage of allowing for static comparison of technology, it has the significant disadvantage of not addressing the operation of the individual technology within the power system as a whole.

9. A gigawatt is 1,000 megawatts, or 1,000,000 kilowatts.

## REFERENCES

Fisher, J. C. 1974. *Energy Crisis in Perspective.* New York: John Wiley.

———. 1978. "Size Dependent Performance of Subcritical Fossil Generating Units." Palo Alto, CA: EPRI Report, October.

Gleick, James. 1987. *Chaos: Making a New Science.* New York: Viking.

Haskell, Elizabeth H. 1982. *The Politics of Clean Air: EPA Standards for Coal Burning Power Plants.* New York: Praeger Special Studies.
Lee, Thomas H. 1976. "The Case of Evolutionary Optimization." Oak Ridge Conference of Association Universities, October 20–21.
———. 1987. "Assessments and Measurements for Energy Conversion Systems." International Institute for Applied Systems Analysis. Working paper number WP-87-27. Laxenburg, Austria.
U.S. Atomic Energy Commission. 1974. *Power Plant Capital Costs, Current Trends and Sensitivity to Economic Parameters.* Washington, DC.

# Part II
# Lessons

# 5 What Went Wrong?

What went wrong? How did we manage to shoot ourselves in the foot? Should we attempt to analyze all the individual blunders, and in the process identify the key players, their interests, and the environment in which the blunders were committed? We will not do that because our intention is not to point the finger of blame. We do not subscribe to any conspiratory hypothesis; there are no culprits to identify and punish. The energy leaders and experts have already paid dearly. They have lost personal and corporate fortunes and prestige, all with the best of intentions. Instead, we will discuss the causes of the blunders, for many are, in a sense, generic. They are generic because blunders in fields other than energy have been caused by them as well. Specifically, we will discuss five areas, with examples.

The five subjects are: mistaken assessment of the value of quality, inadequate measurement systems, lack of a systems approach, misapplication of models and simplistic assumptions, and misunderstanding the dynamics of technology.

## THE MISTAKEN ASSESSMENT OF THE VALUE OF QUALITY

Quality has been a subject of intensive debate under the general problem of international competitiveness (e.g., Ishikawa, 1985, and Deming, 1982). One may wonder whether we are beating a dead horse by raising the issue again in this book. We believe not.

We bring up the issue of quality principally because some of the blunders in the energy field were caused by mistaken assessment of quality. A notable example is the case of commercial nuclear power. The same philosophy—overemphasizing the importance of near-term financial results over quality—that caused the problems with nuclear power is now recognized broadly as a reason for the loss of U.S. competitiveness in other industries.

Much of the current discussion on quality as a root cause of the loss of U.S. international competitiveness has dealt with management systems and philosophy. There is a tendency to blame American managers for a shortsighted outlook and dependence on outmoded management practices that led to an inadequate emphasis on product quality. We believe these are merely the symptoms, not the cause. The fact that most efforts to introduce Japanese quality circles and total quality control into the United States have not been successful indicates deeper problems. We hope that by addressing the question of quality on a general and somewhat philosophical level, we might uncover some of the problems.

Quality can be roughly divided into three categories: economics, reliability, and robustness. Economics includes issues that are difficult to measure accurately, despite their explicitly quantitative nature—cost per unit of output, efficiency, maintenance costs, product life, service factor, and so forth. Reliability deals with disruptions to the smooth operation of an energy system or its components. Robustness deals with the ability of a system to respond in the face of future uncertainties; for energy systems, this must include uncertainties in the nature and price of the fuel supply. All three categories of quality must be considered in any cost-benefit analysis, along with issues that are important but difficult to measure and difficult or impossible to quantify.

Although the word *quality* may seem ill defined, it is a phenomenon with great vitality and high practical repercussions. A misjudgment in assessing one or more of the three elements of quality can lead to egregious and expensive errors, whose effects might be felt far beyond the confines of the system or firm itself. The difficulty lies in assessing the *values* of these elements, which are not easily quantified. Part of the reason for the difficulty is that the up-front *cost* of achieving quality is both quantifiable and immediately felt, but the *benefits* may be both external and long term—and therefore difficult to quantify in anything but intuitive terms. As the foregoing chapters have made clear, our record of decisions about energy systems in the past fifteen years has been less than glorious. In tackling the question of the role of quality in these decisions, past and future, let us first examine the problem.

If you saw a magazine advertisement for shoes, and if its header read in bold red letters "Quality Pays," you might suspect that you were about to read a sales pitch for a pair of well-crafted but overpriced shoes, whose principal attraction was their brand name rather than their intrinsic utility or durability. Gone are the days when ads carried the quaint legend, "Quality adequate for price charged." Gone also are those helpful Sears' catalogs where the same item was sold in three different qualities, each clearly labeled "good," "better," or "best."

Quality has been sold short in the Western industrialized world. A major oil company promoted its highest-priced motor oil as "the world's finest." It knew that such a vague claim could not be explicitly disputed, whereas, a claim for something specific, like the highest-viscosity index, might be. In this decade, we have seen where the neglect of quality has gotten us, now that we must compete with nations that take quality seriously. The Japanese, once associated with paper parasols and matchstick-strong toy cars, now produce electronic goods and compact cars whose brand names are associated with high quality. And these products have captured an impressive share of our markets.

Quality is a difficult subject to talk about because it is difficult to describe or define (Garvin, 1988). Many say that they "know quality when they see it." Quality is difficult to quantify; and in a left-brained culture that honors analysis over intuition, what cannot be quantified is not taken seriously. This is partly why decisions involving quality are difficult to make.

A rational person, whether manufacturer or buyer, needs two things in making a decision to manufacture or to buy: first, relevant and reliable data; second, workable frameworks with which to evaluate alternatives. Only then can that person make the best choices. But "best" embraces many dimensions. Some of these lend themselves only poorly to quantification or objective evaluation. Best choices involve considering the trade-offs well known in engineering economics. These include such dimensions as capital, operating, and maintenance costs, reliability, economic life, the time value of money, risks and uncertainties, opportunity cost, inflation, and so forth. They also involve that fragile and elusive element known as judgment.

Past decisions involving quality as they applied to energy systems have been frustrated by the use of misleading information, the exercise of inappropriate assumptions, the application of inadequate criteria, the employment of deficient analytical processes, and—yes—bad judgment. But all that wasn't the problem. The problem was, it was not known at the time that the information was misleading, that the assumptions were inappropriate, that the criteria were inadequate, that the processes were deficient, or that the judgments were bad.

Clearly, without the input of sound and accurate data into a decision involving quality, there is little hope an economic, robust, and reliable energy system will emerge.

## Criteria and Context

During the past twenty years the Western world has become a producer of disposable goods, all to achieve the ends of increased efficiency, economy, and functionality. Borstein (1973) suggests that this came about largely as a result of Alfred P. Sloan's strategy for General Motors against Ford before World War II. (See also O'Keefe, 1985.) Ford's dream was to build a car that would last forever, and for every American family to have one. Sloan perceived that GM's need was to keep the production lines running forever, and Ford's strategy would not accomplish that. He invented the "model change," leading to what we now call planned obsolescence. This has reached its logical conclusion in the "throw-away economy."

Although the market has bought the model change and the throw-away economy, these habits may well not be the result of authentic analysis. The market may at some point come to see that it is not an appropriate basis for the production and distribution of goods. At the very least, the market will recognize that the philosophy did put quality issues onto the back burner.

For example, homes represent by far the largest single purchase most people make in their lifetimes. But resale markets for homes have all but eliminated the original owner's stake in investing in long-term quality. The typical, nomadic American who purchases a house knowing he probably will be moving on in a few years has no incentive to pay a premium for efficient heating and cooling systems, insulation, and weatherproofing that would be cost effective. He or she might be the loser in financial terms when trying to recoup the premium at the time of resale. In addition, the homeowner faces an institutional barrier because mortgage institutions tend to lend on the basis of size rather than quality. It is difficult for a prospective buyer to obtain a larger loan because a house contains a superior—but more expensive—heating system. Builders, knowing this, do not incorporate quality to begin with—again, *regardless of the cost effectiveness of such qualities* to all of the owners combined or to the economy and society as a whole.

The point is that appropriate market decisions cannot be expected unless both their context and their criteria are appropriate. If market valuations do not give full credit for the intrinsic value of quality, then quality will not be there. Cultural and institutional barriers such as those described interfere with the appropriateness of context and criteria.

A spectacular contrast is provided by the whole attitude of "Japan, Inc." During World War II, American industrial strategists pioneered new methods of quality control for wartime industries, with the object of producing large quantities of high-quality armaments using an unskilled labor force. Yet, after the war, many of the strategies were not adopted by American industries, despite their proven success.

Halfway around the world, post–World War II Japan faced a completely different situation. Its economy was ruined, and it had to rebuild its industries from the ground up. In 1950, the Japanese Union of Science and Engineering invited Edward Deming to educate them about U.S. methods of quality control. After studying the Japanese work force and industrial structure and operating methods, Deming (1982) decided that a new management philosophy based on the experience of wartime U.S. industries could be developed and implemented in Japan. Accordingly, he met with forty-five Japanese industrialists, and from that encounter, a new management philosophy for Japan was born: quality pays.

The philosophy is not a mechanistic approach to product quality. It is a human approach. The quality circle approach involves workers at all levels. The approach is now well known and is practiced all over the world, including socialist countries. Recently an extension of the concept to a total management system has been introduced. This is known as the total quality control system. It treats innovation management in a way similar to the control of product quality. There are two circles: one for products and one for innovation. The entire organization is involved in a human way to tackle all problems that a corporation must face. Quality is not only a product measure; it is a measure of corporate performance, including innovation (White and Dunlevy, 1986).

Although the systems worked well in Japan, their adaptation in the United States and other countries has not been as successful as expected. Before examining the underlying causes, let us look at a few examples.

**Nuclear power.** Later in this chapter we will discuss the strategic errors of the industry in adopting the economy of scale and market-share strategies without a complete understanding of the necessary conditions for the success of either. We point out that a strategy based upon high quality is much more conservative and robust with respect to uncertainties. The neglect of the importance of quality by the nuclear industry is not only strategic but also painfully obvious operationally. Although the planning assumption for the capacity factor of nuclear power plants was 70% in most cases, the actual performance in the 1970s was less than 60%. Examining the causes of loss in capacity factor for either BWR or PWR reactors, we find that most of the causes were not nuclear, that is, not associated with the fission reaction. Failure of "traditional" equipment from inadequate testing of valves, lack of attention to the possibility of stress corrosion cracking, and flow-induced vibration, and underestimating the importance of water chemistry were responsible for the lion's share of the loss in capacity factor. These are problems related to mature technologies. Their failures cannot be attributed to anything but poor attention to quality. A review of what is now public knowledge illustrates the point vividly.

In the early 1970s, twenty-five years after the beginning of nuclear power, General Electric reexamined its nuclear strategy. Serious questions were raised on the viability of the learning-curve and market-share strategy. Management felt that a quality strategy might make much more sense, since the backlog was in the order of 60 gigawatts, which amounted to several billions of dollars. If the plants did not perform to their expected levels, the financial liabilities could be astronomical. A task force was appointed by the then-chairman of the company, Reginald Jones; a senior vice president, Dr. Charles Reed, headed the study. The conclusions of the internal Reed task force, ultimately reported by Wolfe (1987), though they concentrated on BWR only shed a great deal of light on how the industry as a whole treated the quality issue. In the Reed report, concerns were raised on the realism of the plant availability goals, the adequacy of design margins, the overall approach to design and development of plant reliability, and the need for testing facilities for the performance of important components. According to Reed as reported by Wolfe:

> The Nuclear Energy Division (NED) design program to improve Reliability/Availability/Maintainability, outside of the safety area, is largely undefined and unimplemented. Action in this area is not yet consistent with a BWR quality leadership strategy. . . . The plant availability goals which have been set seem unlikely to be met. These goals have not been adequately evaluated statistically to assure they will provide the BWR a competitive edge in reliability/availability/maintainability.
>
> Some of the engineering design margins are considered insufficient to ensure full power operation reliably through plant life. . . . Additional facilities are needed to assure full scale life testing and design qualifi-

cation prior to installation and operation of certain important plant components.

A recommendation was made to "embrace the discipline of consistently providing for the resources and time cycle needed to accomplish adequate design qualification and endurance testing before making a decision to include in a commercial offering a new unknown hardware design which is important to plant reliability." (Wolfe, 1987, pp. 256–372.)

As a consequence of the Reed study, GE did construct a number of new test facilities, improve the reliability of some components, and institute a host of other changes. To accomplish them, management permitted the business to continue operating at a loss. As a result, today the performance of BWR-6 plants compares favorably with the best reactor types throughout the world. Nevertheless, the neglect of the quality issue by the entire nuclear industry had far more serious consequences than the competitiveness of BWR versus PWR. What was lost was public confidence in nuclear power.

The Reed study was not a safety review. But one can hardly ignore the safety issue in discussing quality. Certainly, the slow death of the nuclear power business in the United States has had a great deal to do with the debate on safety. In retrospect, one cannot help believing that the Atomic Energy Commission's safety R&D program was inadequate for the job. The AEC strategy of allowing simultaneous development, testing, and commercial deployment and inattention to the problems of the fuel cycle contributed significantly to bringing the nuclear industry to where it is today.

**The combined cycle system.** That generating efficiencies can be greatly improved by combined cycle systems has been known for a long time. In a gas-turbine combined cycle system (GTCC), the exhaust from a gas turbine is fed into a residual heat boiler, which generates steam for a steam turbine, which drives yet another generator. Energy that would have been sent up the stack is captured in this "bottoming cycle." If natural gas is used to fuel the gas turbine, the overall efficiency can be higher than 50%, compared to the 30% to 35% efficiency of conventional coal, oil, or nuclear plants. The capital cost of such a system is in the range of $500 to $750 per kilowatt, compared to the more than $1,000 per kilowatt for coal-fired plants. Yet, its adoption as an important option for power generation was late in coming. The principal reason was lack of quality in manufacture, and this in turn was reflected in the downtime for the GTCC.

The reason was easy to understand. Gas turbines were only used for peaking duties, such as the air-conditioning load on summer days, which require additional generating capacity a few hours a day. Because of the nature of their application and the relatively short time required for maintenance, their availability and reliability were of less concern to the power industry. Yet, when the gas turbine combined cycle was introduced for midrange and base load applications where the demand of electricity lasts for a large part (midrange) or 100% (base load) of the twenty-four hours a day, the same types of gas turbines were used. In addition, the control system for the combined cycle system did not receive as much attention as it should have. The result was miserable operational performances. But the problems that caused the poor reliability were clearly solvable. Another type of gas turbine—the jet engine—was capable of running 50,000 to 100,000 hours without repair. The technology to build reliable control systems was also well known. If sufficient effort had been applied to improve the reliability of combined cycle systems, the probability of success would have been very high. But it was not done and the acceptance of this important technology was therefore delayed.

## Underlying Structural Causes of Low Quality

There are three underlying causes of low quality: faulty accountability systems (net present value and life cycle cost), structural decentralization, and institutional barriers.

**Faulty accountability systems: net present value.** The classic criticisms of the use of net present value and other traditional analytic systems are especially relevant to the quality issue. They are concerned with the mathematics of financial analysis, which exponentially diminishes the impact of all factors as they extend into the future. Readers familiar with net present value (NPV) analysis will recognize its operating formula:

$$NPV = CF_o + \frac{CF_1}{(1+i)^1} + \frac{CF_2}{(1+i)^2} + \cdots + \frac{CF_n}{(1+i)^n}$$

where $CF$ is cash flow, $i$ is the firm's required rate of return ("hurdle rate"), and $n$ is the project life in years.

In normal usage, a manager contemplating a new investment in cash-flow-producing equipment figures the initial purchase cost to be $CF_o$ (making $CF_o$ a negative number), then estimates cash flows generated by the equipment for a number of years into the future. If the firm requires a 15% rate of return from new investments, he or she will set $i$ to 15%. If the NPV calculates to be zero over the project life, the new investment will be estimated to meet the firm's rate of return; if more than zero, that's even better.

As one can see, the cash flows generated in future years have less and less impact on NPV as the years go by. At commonly used rates of return, benefits beyond the fifteenth year have little impact, and those beyond twenty years hardly show up at all. Since many of the benefits of quality are future oriented, their contribution to NPV analysis is greatly impaired.

This short-term view, caused by the *form* of analysis, is exacerbated by the custom of requiring mobility among managers and executives, who therefore do not remain in the same position long enough to experience the long-term benefits real quality can provide. Not only is it difficult for the decentralized manager to justify paying the price for the quality he or she knows should be bought, but the manager probably will not be around when the decision is proven right.

Another problem is the exclusion of costs and benefits that are external to the decision maker. Social costs and benefits are excluded, as well as very long-term and intangible benefits, even to the firm. Realistic and informed decision making requires an accountability system that includes external factors.

The decentralization of decision making in American firms has exposed them to two other dangers. The first is the tendency of first-line managers to gold-plate, and therefore the need for higher-level management to apply downward pressure on expenditures. The unintended result of the pressure is a reduction in quality.

Without the proper constraints, there is a natural temptation for the *user* to make things nicer or more convenient than they *need* to be. The temptation is not limited to a district manager selecting opulent office furnishings, although this is an excellent example. The problem extends to a refinery manager specifying shorter, more easily climbed tanks instead of taller, more economical ones, or specifying stairs when ladders would do, and so forth. But the difference between gold-plating and functionally desirable quality is often difficult to spell out, the procedural protection against gold-plating having made it difficult to make the added expenditures that are in fact justified for additional quality.

The second danger is possible underspending by buyers and underpricing by suppliers. Shopping for the bargain, or pricing competitively, requires effort and skill. But such widely accepted protective policies as buying against specifications and competitive bidding assure not *appropriate* quality, but rather *minimum* though, we hope, adequate quality. We are reminded of a friend's visit one rainy day to a newly constructed state university building. Noting many buckets catching roof leaks, inoperable door-latches, and other defects, he asked who built the place. "The low-cost bidder," was the answer. Bidding practices are heavily weighted to protect the firm against downside risk—against paying too much, against Type I[1] errors. Seldom is there concern over Type II[2] errors: paying too *little*. A *larger* expenditure almost always must be justified; but a *saving* is almost always unquestioningly perceived by top management as praiseworthy. Managers are not hauled before the board of directors for underspending the budget. Such tendencies mitigate against *quality*.

Before we leave the issue of net present value, we should touch on the subject of vision and judgment. A quality-related decision pans out only after many years, and quarterly reports be damned. As in a chess match, the midgame status of pieces lost and won may have little direct relation to the final checkmate. The fact is that it often is not until the long run that quality bears fruit. For a firm to make *good* quality decisions, it must have a broad *vision*.

Many of humankind's treasured accomplishments never would have survived an NPV analysis. One of the seven wonders of the ancient world—the Colossus at Rhodes—didn't; it was melted down so the bronze could be used to make cannons. There is no NPV rationale for the Roman aqueducts of Southern Europe to have been built so well that they are still in use today. The 450-year old El Morro in San Juan, Puerto Rico, should have been built so that it fell apart at least 200 years ago; yet, even now it appears as if it might serve its original purpose.

In decision making, there is no substitute for vision and judgment. NPV can be helpful in protecting a decentralized organization against gold-plating and overspending, and it provides a rational yardstick for measuring and comparing new projects. But, like any tool in the wrong hands, it can be dangerous. Improperly applied, NPV can inhibit the vision, foresight, and judgment that allow quality to receive just consideration.

**Faulty accountability: life cycle costing.** The concept of life cycle costing has all the appearances of being as simple and practical as its name; unfortunately, it is most often honored in the breach. Life cycle costing can be straightforward, if we would only be guided by it. But it is difficult to give life cycle costing the attention it deserves when the "life" extends beyond our tenure, or beyond the time our assumed discount rate allows us to consider as relevant.

Life cycle costing simply means taking into account *all* costs over the life of the system: initial cost, operating and maintenance costs, direct and indirect costs, and salvage values. A popular step was taken toward this method, for instance, with the labeling of the energy usage of refrigerators and air conditioners in the middle 1970s. The purchase price of an air conditioner is only a small element of its total cost. In order to choose rationally among air conditioners, a buyer must be able to estimate operating costs over the complete life cycle of the unit. And it is over the life cycle that quality usually carries the day.

The point is that a life cycle cost analysis—although not particularly esoteric or difficult to perform—can yield results that are not obvious and tend to bring decisions down on the side of quality.

**Structural decentralization.** Structural decentralization is the second reason for the neglect of quality. The concept of decentralization was introduced into GE in the early 1950s. Responsibility, authority, and accountability were subdivided among the managers for specific operations within the company. Thus, managers were given great authority in how they conducted their part of the business and made decisions, and were held accountable to central management for their results.

GE recognized that the method for measuring results was critical. Accordingly, the company developed a system to measure them. The system consisted of eight separate measurements (see Table 5-1),

**TABLE 5-1 Key Result Areas and Performance Measurements Originally Defined by General Electric**

| *Area* | *Performance Measurement* |
|---|---|
| 1. Profitability | Residual Income |
| 2. Market Position | Market Share |
| 3. Productivity | Output (Value Added)/Input (Payroll + Depreciation) |
| 4. Product Leadership | —Competitive standing<br>—R&D and innovation in Engineering, Manufacturing, and Marketing |
| 5. Personnel Development | —Development programs<br>—Inventory of promotable people<br>—Effectiveness of program implementation |
| 6. Employee Attitudes | —Statistical indicators<br>—Periodic employee surveys |
| 7. Public Responsibility | Surveys |
| 8. Balance between Short- and Long-range Goals | All previous factors have short- and long-term dimensions |

*Source:* Adapted from Robert N. Anthony and Robert H. Caplan, "General Electric Company," 9-113-121. Boston: Harvard Business School, 1964.

from near-term profits to the balance between long- and short-range goals. Near-term profits were quantifiable; the balance between long- and short-range goals was not. The relative weight given to the measurements forced managers to devote most of their attention to achieving the near-term goals. Quality was not explicitly mentioned as a key area or measurement of performance.

Decentralization at GE was based on two other major tenets:

- Managers are professionals, in the sense that a competent manager will succeed, whatever the business. Thus, the successful manager of a transformer business should be able to become a successful manager in the nuclear business.
- Managers must have both intrafirm mobility and a broad diversity of skills. The most efficient way to train people is to acquaint them with a broad variety of business functions, to facilitate their ability to move from one function to another and from one business to another.

The consequences of applying these concepts were obvious. General managers moved from one business area to another at a rapid rate. In the late 1950s, the average "half-life" of a general manager's assignment was about two and a half years.

The pressure to achieve near-term, profit-oriented goals was so great that general managers in some decentralized businesses were evaluated *monthly* on sales. Thus, the last week in any given month was exceedingly hectic as efforts were made to ship products to customers, sometimes with lamentable disregard for quality. Shipments were made with important parts "to follow later." Testing was incomplete, with the thought that customers' complaints could be addressed later. As a result, savvy customers learned not to accept equipment shipped in the last week of any given month.

The U.S. financial community was—and continues to be—a key source of pressure for achieving near-term profit-oriented goals. Corporate profits are measured quarterly. The price-to-earnings ratio of the firm's stock depends on its quarterly performance, and so does the financial community's willingness to provide capital.

Given the pressures, managers were driven to gain short-term profits; quality considerations were correspondingly downplayed. Total quality cost was considered, but the internal preventive quality cost was treated as equal to complaint costs. A dollar is a dollar! Whether or not to correct a low-quality feature in a product depended on the cost of doing so, not on potential future costs or liabilities, or even on the loss of customers.

In the General Electric Company, two divisions within the same group had very different strategies: the Nuclear Division pursued the economy-of-scale strategy while the Large Steam Turbine Division pursued the quality strategy. Whenever the two strategies conflicted and headquarters had to resolve the difference, the Nuclear Division usually won. It is easy to understand why. In the years when dreams of the potential of nuclear power were sweet and exciting, market share was the only factor that counted. The price of GE or Westinghouse stock would respond to the winning or losing of an order. Whether the order would eventually lead to profits was not even a concern. Where does that leave quality?

In energy-related industries, particularly electric power, the quality of a manufactured product has always been considered important, but quality was more often used as a marketing tool (high reliability or low maintenance) than as a core component of management strategy. There is an exception, however.

The U.S. electric utility's need to reduce the cost of electricity has always been of paramount importance to managers, customers, and regulators alike. A great deal of emphasis has therefore been placed on improved conversion efficiencies and economies of scale.

In 1970, there were only two dominant suppliers of turbine generators in the United States, GE and Westinghouse, and each pursued different strategies in designing and manufacturing turbines. GE emphasized reliability. Westinghouse emphasized lower cost, considered in the more conventional way. In pursuing the reliability strategy, GE adopted several fundamental policies.

- It decided not to increase the size of new turbines by more than 20% over the largest unit then in operation;
- It initiated field programs to document the performance, failure rates, and failure effects of its turbines in operating plants;
- It assigned reliability-oriented engineers to the field;
- It undertook massive investments in materials R&D; and
- It devoted considerable internal energy to analyzing and promoting the economic value of the reliability approach.

In marketing its turbines, GE pointed out that its customers could not afford to install a competitor's turbines if its downtime for repairs was higher than GE's by even a few percent, *even if the turbines were free.*

In contrast, Westinghouse invested in new design to reduce the initial price of the turbine-generators. Further, it was generally more aggressive in developing larger-sized units, and sought leadership in conversion efficiency.

One interesting feature of the turbine-generator market has always been that electric utilities evaluate reliability and efficiency along with first cost in their purchase decisions, although they must depend in large part on manufacturers' claims in doing so. In the long run, the outcome of the price-quality trade-off was clear. The competitor with the reliability strategy won, as reflected by the market share it has enjoyed.

**Institutional barriers.** Regulation is an inevitable part of modern life, either in the form of legislation enforced by regulatory bodies, or in the form of standards and codes established by professional societies or industrial organizations. When regulations are designed and implemented properly, they can enhance the quality of a system. On the other hand, they can be an obstacle to quality.

Civil aviation is a remarkable success story. It has achieved a safety and reliability (i.e., high-quality) record that society has found acceptable. The same cannot be said about nuclear regulation. Many of the regulatory efforts were directed toward theoretical calculations (and experimental verification, to a limited degree) for very low-probability events that could lead to a disaster. Although these are necessary, insufficient effort was directed to component testing, operator training, man-machine interface, and other aspects that are vital to the reliability of the plant. After the accident at Three Mile Island, overregulation occupied most nuclear reactor organizations to the degree that very few resources were actually available for improving the reliability of plants.

According to the Carter-appointed Kemeny commission report:

> Once regulations become as voluminous and complex as those regulations now in place, they can serve as a negative factor in nuclear safety. The regulations are so complex that immense efforts are required by the utility, by its supplier, and by the NRC to assure the regulations are complied with. The satisfaction of regulatory requirements is equated with safety. (Commission, 1979.)

In this situation regulation had little to do with safety or reliability. After the Three Mile Island disaster, for example, engineers in the nuclear supplier companies had to spend so much time satisfying the regulatory requirements that they hardly had time to work on the reliability of the plants. Thus, it is not unreasonable to say that in this case, regulation *impeded* quality. The present situation in nuclear regulation requires major surgery, not a bandage. The Kemeny commission recommended a number of improvements such as definition of the mission, clarification of the relevance of tasks, simplification and clarification of procedures, and also possibly major reorganization. The blunders have already been committed, but corrective efforts are nowhere in sight.

At times, tradition replaces logic and pragmatism in quality considerations. Building and construction codes, as well as professional and trade practices, are examples. The codes and practices are especially restrictive in the face of rapid technological advances. In constructing a building, better (i.e., higher-quality) may be cheaper, but the means of achieving it may be illegal or beyond consideration because of building and construction codes. Technological advances are constrained only by the human mind, but changes in codes and practices must wait for the wheels of bureaucracy to turn. A new, cost-effective insulation technology can take years to gain acceptance, especially if it disturbs established craft techniques or distribution channels, or if it changes the labor content.

Until a decade or two ago, the sheer technological dominance of the United States masked our inattention to and lack of appreciation for quality. Where quality did show, it was often a result of doing what was technologically possible.[3] The quality issue, especially as applied to energy, was largely unappreciated by our culture, underrated by our systems, undervalued by our analyses, misunderstood by our criteria, and ignored by our institutions.

Losing market share to Japan in many fields awakened us to the importance of quality. Whether our system or our culture will permit us to correct the deficiencies, only time will tell. The issue is not purely a technological one. A recent study by Jaikumar (1986) pointed out that we are losing the battle in computer-integrated (so-called flexible) manufacturing to Japan—not because of lack of technology, but because of inadequacies in organization and management competence.

The Komatsu case in Japan is instructive. When numerical control (NC) lathes were introduced, the security of the skilled labor force in the company was threatened. But through the efforts of management, the workers were led to understand the importance of this advance and many skilled workers turned themselves into experienced NC programmers. This happened again when robots were introduced three years later. The need for top managers to appreciate the virtues of quality is now critical.

## INADEQUATE MEASUREMENT SYSTEMS

In many ways, the problem of wrong measurements is similar to the problem of wrong assumptions. Both have a tendency to stick. Perhaps it is human nature. No one likes to admit mistakes, so we stay with our assumptions and measurements even though they turn out to be wrong. We use them to justify our actions. Nevertheless, the problems associated with measurements may be more serious and difficult to crack than those of assumptions, because there may be infrastructure barriers associated with measurements that are not present with assumptions.

We will examine two areas in this section: measurement problems in the electric industry and the problems caused by financial measurements in the United States.

### Measurement Problems in the Electric Industry

The cost of electricity (COE) has traditionally been the most important measure of relative merit for generators in the electric power industry. The cost has three components: capital, fuel, and operation and maintenance (O&M). All three are easily understandable, and how they are calculated is traditional. So, what can go wrong?

Consider a single generating unit, a nuclear or coal unit. We understand its capital cost, we understand the amount of fuel that it will take to produce a single kWh, and we think we understand its O&M characteristics. All that we need is an estimate of the number of hours that the plant will operate during an average year and its likely lifetime and we can calculate the COE. Where is the fallacy? One fallacy is that no plant operates alone and the number of hours of operation is determined by the *system* conditions. It is the output of systems analysis, not the input.

There is another fallacy: Consider a utility company that owns a nuclear power plant. It is on-line and in the rate base, and is one of twenty or thirty generators. In daily operation, the dispatcher in the company must decide, based on the forecast of demand for that day, which generator to use. Because the capital costs are already in the rate base, the dispatcher only considers the marginal cost, which is the sum of the fuel cost and the O&M cost where fuel costs dominate.[4]

The same company has another plant with much lower capital cost but somewhat higher fuel cost. Such a plant may be a gas-fired combined cycle plant.[5] The correct decision for the dispatcher is not to run the combined cycle plant because the marginal cost is too high.

But, in planning, a different criterion—least long-term cost—is used. In this criterion, the capital cost is a very important factor in the optimization equation. Let us see how this criterion will influence the "fate" of combined cycle systems.

For years, combined cycle systems were never considered seriously in generation planning. There are two reasons. First, the early combined cycles were not reliable, that is, they were perceived as low-quality generators. Second, there was a significant concern for the price and availability of the fuel. Given these reasons, the alternative was simply not being considered in capacity-planning decisions, even though, as we pointed out before, the overall efficiency for commercial gas-fired GTCC systems can be as high as 50%. The capital cost of such a system is about $500/kW (1988 dollars).

Two factors brought about a change in the situation. First, the Electric Power Research Institute (EPRI), in cooperation with Southern California Edison and Texaco, proposed to organize a consortium to support a $300 million dollar project to develop an integrated coal gasification combined cycle system (the Coolwater Project). GE joined the consortium with responsibility for the combined cycle portion of the system. For the first time, the reliability issue of combined cycle systems was addressed in detail.

Second, Japan, in an effort to diversify its energy sources, ordered several gigawatts of combined cycle systems using liquefied natural gas. Suddenly, combined cycle systems came into the spotlight.

At the Massachusetts Institute of Technology (MIT), a careful systems approach was taken to look at the potential of combined cycle systems in future generation planning, a subject that has received little attention so far (Tabors and Flagg, 1986). The analytical system used for the analysis was the then newly released electric generation expansion analysis system (EGEAS) (Caramanis, Tabors, and Schweppe, 1982) developed by MIT and Stone and Webster Engineering Corporation for EPRI. The study was carried out for a fifteen-year planning horizon beginning January 1, 1990, and ending December 31, 2004.

With a set of reasonable assumptions, the study concluded that natural gas fired combined cycle systems offer an important alternative for capacity planning, a surprising conclusion to many planners at the time.

The authors then proceeded to forecast the impact of likely technical advances of the combined cycle system. The development of gas turbines was heavily influenced by jet engine technology. The U.S. government, over several decades, invested hundreds of millions of dollars annually to improve the performance of jet engines. For aviation applications, size and weight are significant constraints. The same is not true for ground applications. Thus, an optimum design for aviation application may not be the optimum for electricity generation. An optimum design for a freestanding gas turbine may not be the optimum for combined cycle systems. As amazing as it may seem, this fact was recognized only very recently. There is now general agreement that if reheat is introduced between combustion stages in a gas turbine, the turbine can be optimized for combined cycle systems. A number of ideas on reheat gas turbines have been proposed (Lee, 1988). In our laboratory in MIT, El Masri showed that efficiencies between 55% and 60% are achievable for reheat combined cycle systems. (El Masri, 1985, 1986.)

When the possibility of high-efficiency combined cycle systems was included in the EGEAS study, the results were indeed remarkable: combined cycle systems become the dominant generation option.

Combined cycle systems make economic sense for future generation additions; from a systems perspective and given our understanding of the likely cost of fuel, the electricity generated by them is lower in cost. From the perspective of the dispatcher, once on-line, they may not be the most economical to run. If fuel prices increase, the plants will be moved higher and higher in the loading order and run less and less. Our short-term criterion (least marginal cost) and our long-term criterion (least long-term cost) are different.

We are faced, therefore, with conflicting measures for planning and for operations. Interestingly, in the recent push for generation by independent power producers, the contractual criteria imposed by regulators (almost a COE) are more like those used by planners than by dispatchers. Therefore, because of long-term contracts, if a third-party generator enters the market as a base load supply, it will be treated as such, regardless of the economics of the dispatch logic. By changing the ownership from the third party to the utility company, the same system would cease to be dispatched as base load if the price of the fuel increases significantly. Isn't this absurd?

So far we have only dealt with measurements for well-established technologies. For new technologies, the single measurement—cost of electricity—is not only useless, but can be outright dangerous. Let's look at the early 1970s when the country was trying to find ways to save scarce resources. Numerous ideas were proposed for energy conversion systems. They included fluidized bed combustion, combined cycle systems, and magnetohydrodynamic systems. Again, COE was the most important measurement and contributed to poor decisions. On capital cost, we cannot even have good numbers for established technologies like coal-fired plants (see Chapter 4). How much confidence dare we have in the capital costs of plants utilizing untried and unproven technologies?

On fuel cost, the fallacy is even worse. If one of the potential features of the new idea is higher conversion efficiency, of course the fuel cost would be assumed to be lower—a built-in advantage for the new idea, whether it is true or not.

Operation and maintenance costs are usually assumed to be insignificant in the conventional approach because of the assumption of constant capacity factor. That cannot possibly be true for plants with new technologies. As a matter of fact, the actual O&M costs may well be the most important component in the calculation of COE. Witness the nuclear case: the most profitable segment of the nuclear industry is the services business—just to keep the plant running.

Nevertheless, if we consider the past, from the days of introducing nuclear power to the mid-1970s, COE was always the dominant measurement. The deficiency of COE as the dominant criterion was recognized in the mid-1970s. We mentioned in Chapter 4 in a study sponsored by NASA and subsequently by EPRI that two additional methods were proposed to complement the COE calculations. The first one was the more conventional present worth calculations. For that calculation, one has to estimate:

- the total R&D cost over time;
- the year when the new technology would start to penetrate the market;
- the likely competing technologies at the time of its introduction; and
- the savings from its introduction compared to the case without it.

There are fundamental deficiencies in the method, aside from the obvious uncertainties:

- Penetration in the market was assumed to be determined by economics in the traditional optimized generation planning calculations; that is, the penetration is falsely assumed to be instantaneous if economics is in its favor.

- Savings were calculated without considering who benefits and who pays (this was recognized by the researchers).

Considering the uncertainties and the deficiencies in the present worth calculations, one cannot help concluding that the calculation is nearly useless.

The third method proposed was the so-called direct weighting method. Twenty-six intangible variables were identified. By considering the objectives of the utility companies, the researchers proposed a method to calculate the weighting factor for the twenty-six variables. Then, each technological idea was measured against the weighted variables to obtain a figure of merit for that technology.

It was an attempt to reduce our dependence on COE. Our disagreement might be on the statistical weight of such variables as cost of R&D, R&D time required, and the probability of development success. In total, the three variables had a weight of twenty-eight out of a total of one hundred. Some analysts might have given them a much higher weight. But this is a minor point. The real problem with the methodology was that it was too complicated for decision makers. In the face of such sophisticated methodologies, executives found comfort in returning to COE.

The sad thing is that the situation has not really improved. The recent debate on independent power production and on deregulation of power generation brings out this fact. The developers and financial backers of third-party generation rely principally on one measurement: COE. Regulators have found this method comforting because it bolsters their argument: private, competitive generation is less costly. The lesson of the 1970s still has not been completely understood. Forecasts of future costs are always optimistic. It is always easier to choose the less costly forecast even though the uncertainty is far greater. Regulators found it much easier to promote third-party generation based on COE alone than to ask questions such as

- How do we handle transfer of energy across a utility's power lines (*wheeling* in the jargon of the industry)?
- What happens if the transmission system is overloaded?
- What happens when the independent power producer has a forced outage?
- How do we design the *system* for reliability?
- What are the criteria for reliability?

We pose these issues not because they are unsolvable. They are not; they are important measurement issues for power systems. By neglecting them in the debate, we may be marching down the road of simplistic measurements and assumptions again!

### Problems Caused by Financial Criteria in the United States

Early in this chapter we discussed the impact of decentralization and financial measurements on quality. But quality is only one aspect of the measurement problem in U.S. industries.

One of the authors vividly remembers sitting in an executive office, listening to an elegant theory on how the price-to-earnings ratio of the corporate stock is related to annual rate of growth of earnings. The bottom line was, if we wanted to keep the price-to-earnings ratio at 30, the earnings growth rate must be kept at a minimum of 10% per year. That, in fact, became the corporate objective, and the irony is that although the objective was met, the ratio declined.

In 1974, shortly after the oil crisis, cash was very tight in the United States. Even the AAA-rated companies were getting messages from their lenders like, "Don't call us; we will call you." GE was concerned. The vice presidents of finance and planning called a meeting of key managers, pointing out the need to conserve cash, on the order of $500 million in the last quarter. Managers responded. At the end of the year, the reduction in cash usage was 50% better than the corporate goal. The firm's AAA rating was preserved. Very few people asked about the implications of the actions. Were the reductions in inventories going to affect product delivery or product service? Were the pressures on receivables going to have long-term detrimental effects? Or, if these goals were possible, why didn't we do it before? There was one measurement, and there was a single response.

One of the authors managed a high-power testing laboratory for GE. In one year, the business climate was very bad. To save money, he issued an edict eliminating overtime testing unless approved by him personally. One Friday afternoon, he received an urgent telephone call from a department general manager (a high-level position) requesting overtime testing. The laboratory manager knew that department could not afford the extra expenses. He asked, "Why is the testing so urgent?" The answer was, "This is the last weekend in this month. I have to meet my monthly shipment goal." The laboratory manager asked, "What difference does it make at the end of the year?" The answer was, "I don't know whether I will be around then."

When the CEOs are measured by the financial community by the quarter, they are forced to measure their direct reports by the month. The lower-level people may be measured by the week. How do we measure the balance between long-range and short-range interests?

Applying measurements to an operating system is like squeezing a balloon. If you squeeze at one part, the balloon will respond where you squeeze, but other parts pop out. A good measurement system must be one that applies uniform pressure from all sides. Then we can see how strong the balloon is.

## THE LACK OF A SYSTEMS APPROACH

Most problems we deal with fall into a larger context, so that solving one problem often creates another. In the energy field we have too frequently tried to deal with one problem and in the process created another. When we saw uncertainty and potential long-term constraints in our clean energy source (oil), we turned to a dirty but less constrained, indigenous source, coal. The solution seemed simple enough. Was it? Probably not. Increased consumption of coal has clearly been one of the causes of increased acidification and is a major contributor to the greenhouse effect. What we have missed is acknowledging on the front end of our searches for solutions that all things are connected and that some (definitely not all) of the interactions or reactions need to be considered before an action or policy is undertaken. This concern for the larger picture is what we refer to as a systems approach.

In 1965, New York City experienced a blackout. The Consolidated Edison Company lost tens of millions of dollars of revenue. If we had used that figure as the cost of the outage, it would have been completely wrong. The total cost to the city (the system) was estimated to be $350 million, about ten times the cost incurred by the power company. The remaining cost was attributed to lost production, lost time in service sectors such as banking and transportation, and damage to foodstuffs.

At about the same time, Sweden conducted an extensive study of the total cost of power outages in three sectors: industrial, commercial, and residential. Subsequently, the Institute of Electrical and Electronic Engineers (IEEE) made a similar study (Billington et al., 1983). Many follow-on analyses have been made, notably by EPRI.

The conclusions of the studies were similar: the outage cost is much higher to society than to the power company. Now, many years after the studies, we seem to have forgotten the significance of the conclusions. For example, reliability of supply has not been emphasized in the electric power deregulation debate. We do not seem to understand that to avoid the costs of forced outages, we must be willing to invest in preventive measures at the system level. This is not an isolated issue, however. It is a special case of a more general problem: how to assess the total social costs of human development activities. For that we must consider the entire system, which includes not only the individuals, the government, and the corporations, but also the entire ecological system. Unfortunately this is usually not done!

Whenever one discusses the systems approach, confusion reigns. The word *system* means different things to different people. We believe the best definition of a system was that given by Russell Ackoff:

> A system is a set of two or more elements that satisfies the following three conditions: First, the behavior of each element has an effect on the behavior of the whole. Second, the behavior of the elements and their effects on the whole are interdependent. And, third, although subgroups of the elements are formed, each element has an effect on the behavior of the whole and none has an independent effect. (1981, p. 15.)

A system has properties that none of its components has; whenever a component is separated from the system, it loses important properties. The human body is an excellent example of a system. Another example is an electric generation system, with its owners, customers, suppliers, competitors, regulators, and environment. One can never understand a system by looking at the components one by one. One must study the interaction among the components. The social cost issue is a case in point. Another outstanding example is in the environmental area. We now know acid rain is not an isolated problem; neither is the greenhouse effect or the ozone problem. They are interlinked in a complex manner. Only by studying entire systems can we hope to understand such problems and assess likely solutions.

Earlier in this chapter we discussed the relationship between improvements in product quality and the value of the product produced. The conclusion was that for a wide range of products and market areas, quality pays. It often is better to build quality into a product than to repair the product throughout its lifetime. Let us focus now on what may be the corollary.

## Cleaner Is Cheaper

Unlike the previous discussion in which the arguments for quality can be made from within the firm, the discussion of environmental quality must, by its very nature, be outside the firm. The publication of *Tragedy of the Commons* (Hardin, 1968) opened the issue of who benefits and who pays for environmental degradation.[6] The environment has long been considered the ultimate sink for the unwanted residuals of the production process—the flotsam and jetsam of production and consumption. To dispose of wastes into a local river, gaseous and particulate wastes into the atmosphere, and solid and liquid wastes onto the land are ways that an individual firm or a portion of our society exports residuals into the bathtub that is the planet's ecosystem. They represent ways of transferring costs from polluters to the public.

The issue of clean production—in energy, as in other industrial sectors—is a function of who benefits and who pays. If society is indifferent to environmental residuals, the firm will take the economic path of least resistance; it will produce dirty because from its perspective this is the most cost-effective path. If, on the other hand, the cost to society of cleaning up or living with the burdens of degraded surroundings is so great that the costs are pressed back onto the producer, the firm will find it less costly to produce clean.

The issue of who benefits and who pays has assumed great significance as we come to realize that the human race is facing a critical problem: the sustainability of the biosphere.

Alvin Toffler described one side effect of modern industrial society as the rampant, perhaps irreparable damage suffered by the earth's fragile biosphere:

> Never before did any civilization create the means for literally destroying not a city, but a planet. Never did whole oceans face toxification, whole species vanish overnight from the earth as a result of human greed or inadvertence, never did mines scar the earth's surface so savagely; never did hair-spray aerosols deplete the ozone layer, or thermal pollution threaten the planetary climate. (1980, p. 121.)

The issues we face now are not simply economic, such as, How much do sulfur oxide scrubbers increase the cost of electricity? Or, How do we handle the sludge? The issue is, How close are we to reaching the biosphere's ability to absorb humankind's gross insults? Can we respond with technologies that will pull us back from the limit of sustainability? Do the technologies even exist? Yet, whenever we raise these questions, we always end up asking:

- What is the real cost?
- Who will benefit and who will pay?
- How can the costs and benefits be measured?
- How are costs and benefits linked?

Questions like these are currently the foundation of serious debate in Europe and North America. Interestingly, the debates, instead of being on the economic and political issues, concentrate on scientific uncertainties. Consider a few.

*The problem of acid rain:*

- Although the mechanisms behind the widespread destruction of forests are not well understood, the effects of acid rain on soil and water are well established. In Sweden, four thousand lakes have become fishless in the past years, and yet, resistance to controlling acid rain rests mostly on the lack of knowledge on the rate of forest die-off.

  The most compelling aspect of pollution is its effect on human health and well-being. It is easy to imagine the following debates.

*The health effects of sulfur oxides:*

- The argument: Recently the World Health Organization reviewed existing knowledge and defined concentrations for sulfur dioxide and total suspended particulates above which specific effects on health can be observed (World Health Organization, 1978).
- The counterargument: Sulfur dioxide may not be the main agent responsible for the effect. Certain sulfate aerosol particulates in ambient air may be the causal agents (OECD Report, 1981).

*The health effects of sulfates:*

- The argument: Lave and Seskin (1977) found that reducing ambient concentrations of sulfate by 10% was associated with a 0.5% decrease in the total mortality rate across 117 Standard Metropolitan Statistical Areas (SMSAs).
- The counterargument: There is no adequate evidence to indicate the linear dose response relationship. The relationship assumes the absence of a threshold that is simply wrong.

*The effects on farm produce, plants, and forests:*

- An argument: Ozone reduces sugar in grapes as well as reducing the vineyard yields themselves.
- Another argument: Acid deposition dissolves ordinarily insoluble compounds; the resulting salts are toxic to the roots of young trees. The most detailed study on the effect of sulfur dioxide on crops was done for rye grass (OECD Report, 1981). The OECD developed a dose-to-yield relationship for sulfur

dioxide and rye grass, and used that relationship to estimate damage by sulfur dioxide to other crops in Europe.

- The counterargument: There is no evidence that the application of the dose-to-yield relationship to other crops is justified.

One can think of similar arguments and counterarguments on carbon monoxide and heart disease, lead in the blood, and the greenhouse effect.

The debate will continue, and confusion will persist. Whether debates on scientific uncertainties will ever come to a useful conclusion is highly questionable. Alvin Weinberg (1985) pointed out that we may have to live with the reality that some questions may never be answered with the scientific accuracy we like to see. Still, we must make decisions.

Despite the scientific uncertainties, the issue of total social cost is receiving more attention. In the section that follows, we review three sources that discuss the issue. Our objective is reporting; we neither support nor challenge the accuracy of the studies.

A study by OECD (1981) examined in some detail the costs and benefits of controlling sulfur dioxide emission. The study estimates the annual emission for the year 1985 in the absence of control technologies, and then proceeds to examine the costs and benefits of different control scenarios. The scenarios vary from what is likely to be the case in 1985, which includes an agreement by all countries to reduce emission by 37%. The major damage from sulfur dioxide considered in the OECD study was the effects on materials, crops, human health (morbidity rates only), and aquatic ecosystems.

The conclusion was that the costs and benefits are generally of the same order of magnitude; a dollar spent on pollution reduction yields approximately a dollar's worth of benefits. An important point to remember is that mortality rates were not translated into monetary values, although they were translated into life expectancy terms. To gain some insight into the importance of the effect on human mortality rates, the OECD study considered the effect on the lifetime income of increased life expectancy of an average European. When that was included, the benefit outweighs the cost. Of course, the issue of who pays and who benefits remains.

Awad and Veziroglu (1984) made an ambitious attempt to assess the costs of environmental degradation over a broad range:

- Effects on humans: Fossil-fuel-related deaths (51,800 per year), medical expenses related to pollution, and loss of worker efficiency.

- Effects on fresh water sources and resources: Pumping of lime into lakes, loss of fish population, and leakage from gasoline tanks in service stations.
- Effects on farm products, plants, and forests: Ozone responsible for the loss of 5% of annual farm yield, acid rain responsible for additional 5% damage, and acid rain effect on forests.
- Effects on animals: Loss of 10% of total value of livestock and wildlife.
- Effects on buildings: 10,000 equivalent historical buildings (EHBs)[7] each costing $2 million per annum.
- Effects on coasts and beaches: Oil spills from tankers, oil spills from offshore-related accidents, and ballast water discharged from tankers.
- Effects of rising ocean levels: Costs of protecting shorelines, riverbanks, and estuaries; seepage water pumping; and modification of some roads and bridges.
- Carbon dioxide effect on climate.

According to their calculations, the total damage amounted to $7.64 per gigajoule (1984 dollars),[8] or $214 per ton of coal equivalent (tce). The total damage from natural gas is about $140 tce, far less than from coal.

**TABLE 5-2 Estimate of the Damage per Unit to the Society from Thermal Stations Organic Fuel in 1984 U.S. Dollars per tce**

| *Environmental Impact* | *Type of Fuel* | | |
|---|---|---|---|
| | *Hard* | *Liquid* | *Gaseous* |
| *1* | *2* | *3* | *4* |
| 1. Population | 38 | 38 | 13 |
| 2. Stock of land | 15 | — | — |
| 3. Agriculture and forestry including: | 29 | 29 | 9 |
| plant growing | 12 | 12 | 4 |
| cattle breeding | 5 | 5 | 1 |
| forestry, fauna, and flora | 12 | 12 | 4 |
| 4. Fishing resources and fishing industry | 3 | 3 | 1 |
| 5. Buildings and structures | 32 | 30 | 13 |
| 6. World ocean coast | — | 6 | — |
| Total 1–6 | 117/107* | 106/92 | 36/30 |

*Numerator: damage to coastal countries; denominator: damage to inland countries.
*Source:* Thomas H. Lee, editor. 1988. *The Methane Age* (Dordrecht/Boston/London: Kluwer Academic Publishers), p. 159.

Finally, the Working Consultant Group of the President of the Soviet Academy of Sciences on Long-Term Energy Forecasting, under the leadership of Academician M. Styrikovich and others, recently refined the same calculations. (See Table 5-2.) The difference between coal and natural gas in total damage is $60 to $80 tce. The *difference* in total social cost between methane and hard coal is greater than the present market price of coal.

Are these cost figures accurate? Most likely not! Will they be used in making policy and industrial decisions? It is doubtful. Will these effects eventually influence societal decisions? Most likely yes. The societal decisions on technologies are frequently binary and not based on precise quantitative measures.

We have only discussed one issue in examining the problem brought about by the lack of a systems approach: total social cost and, specifically, environmental costs. In practice, failure to consider the total systems in noneconomic areas has caused equally if not more serious consequences. The failure of systems analysts to make decision makers more sensitive to this need is the subject of the next section.

## MISAPPLICATION OF MODELS AND SIMPLISTIC ASSUMPTIONS

Modeling in scientific research is like practicing in sports: it is part of the game and necessary for progress. Newton's laws are models. When they failed to explain observed reality at very high velocities, special relativity—a different model—was introduced. When it failed to work at subatomic levels, quantum mechanics—another model—was introduced.

In general, successful models deal with a small number of quantifiable variables and reliable assumptions. When the number of variables is great and the assumptions are shaky, the model is less accurate and less dependable—but still a model. Modeling socioeconomic systems, unfortunately, falls into the latter category. The number of variables is great, and many are not readily quantifiable. The number of assumptions needed is also great, and many are shaky. For these reasons, efforts at modeling socioeconomic systems have not been as successful as hoped.

The mathematical techniques developed to handle socioeconomic models are often referred to as operations research (OR) techniques. Given the nature of the problem and the constraints, these techniques can find optimum solutions in a number of cases.

When operations research (OR) was first introduced, hopes were high that it would prove to be a new, extremely powerful tool for improving people's capability to manage resources, organizations, and people. OR was expected to provide a better understanding of an increasingly complex society. The results were disappointing, and the glamor of the frontier discipline soon faded. In many industrial organizations, OR lost much of its credibility. Prof. Russell L. Ackoff, one of the first to introduce OR in American academic institutions, said:

> I submit that OR was once a corporate staff function, because corporate executives believed it could be useful to them. It was pushed down because they no longer believed this to be the case, and they correctly perceived that if it had any use, it was in the bowels of the organization, not the head. My observation of a large number of American corporations reveals that when it could no longer be pushed down, it was pushed out. (Ackoff, 1979, p. 93.)

But Ackoff was not the only one who observed this phenomenon. In 1987, one of the authors was invited to give the closing speech at the European OR society's annual meeting. The opening speech by the golden award winner was entitled "The OR Crisis." The issue was one that had become prevalent in the OR community: "Why don't decision makers use our results?" (See also Ball, 1985.)

Nevertheless, among energy experts and within the energy industry of the 1970s, the promise of OR held sway. The temptation was great. We knew that energy was a critical and an extremely complex social-economic problem, and we needed quantified answers quickly. The skills were available to build a complex model that would digest huge quantities of data, and print out huge quantities of "results," usually in the form of multiple scenarios. Shortly after 1973, massive models were built to examine alternative scenarios of demand, supply, and the entire system relating demand to supply. A number of famous models and modelers appeared on the scene. Many institutions were involved in examining and comparing the proliferation of large, sophisticated models. One of the authors chaired one such institution's projects.[9] Eager users spanned government agencies, industrial associations, individual companies, and academic institutions.

Now, fifteen years later we ask, Have large models been useful in solving the world's energy problems? The best one can expect is an answer fraught with hedging and controversy. Why? We believe there were four major problems with the way models were used: first, overindulgence in the optimization techniques; second, a tendency to accept without questioning the quality of assumptions, particularly those underlying large, complex models; third, inadequate attention to the man-model interface, and overreliance on forecasts. We illustrate the problems with a few examples.

### Overindulgence in Optimization Techniques

About twenty years ago, the General Electric Company launched a major effort to apply analytical techniques to plan additions to power generation. The purpose of the program, known as Optimized Generation Planning (OGP), was to find the most economic mix of generation additions among various options such as nuclear, oil, coal, and hydro for power companies.

The U.S. electric utility industry was a pioneer in adopting sophisticated analytical techniques for its operations and planning. Generation dispatching required not only sophisticated mathematical methods but also special computers, built long before digital computers became popular. Thus, one would expect a readiness on the part of utility management to accept the output of OGP.

The initial program was ambitious. Its database included every important generating facility in the country and the load profiles and weather information in different regions over an extended period of time. The goal was to produce a twenty-year expansion plan specifying the most economic mix of generation additions on a year-by-year basis.

Outputs of the initial runs were very interesting. They indicated that the utility companies should have ordered more nuclear power plants and gas turbines. GE took the results to the Tennessee Valley Authority (TVA), the largest utility company in the United States. TVA's initial reaction was to ask why a large utility company like TVA should consider small-capacity gas turbines? GE made a case for gas turbines by pointing out that fuel economy was not very important if the machine was only required to run as a peaking plant for a few hundred hours a year. TVA managers did listen, but with one ear.

For years, GE's analysts continued to use the OGP program for internal planning activities as well as for individual utility companies. They discovered that, for some reason, the utilities always ordered fewer nuclear power plants and gas turbines than called for by OGP, in spite of the overwhelming evidence of economic attractiveness.

In searching for the reasons why utility company buying practices differed from the optimum economic mix, a number of nonquantifiable variables surfaced: utility operators' lack of experience with gas turbines, uncertainty about the acceptability and economics of nuclear power, and their more comfortable feelings about traditional technologies, such as pumped hydro, coal, or oil-fired plants. Thus, GE planners found it necessary to second-guess the output of OGP for their own planning.

The turbulent situation after 1973 made business managers in the electric industry more interested than ever in forecasting the future, and OGP was acepted as an important tool for strategic planning in GE. In 1975, a forecast was produced for the entire United States (see Table 5-3). For this "base case" forecast, a number of asumptions had to be made. They included the capital cost of the generating plants, the future price of fuel, the inflation rates during the planning horizon, and a number of other economic and technical assumptions. The assumptions were supposed to be the best available ones from experts. Hence the name "base case." The forecast called for a total addition of 242 gigawatts of nuclear plants in the succeeding fifteen years (1975–1990). About 90 gigawatts of nuclear plants had already been ordered. With a construction time of about ten years, the industry would have to order 152 gigawatts of new nuclear plants in the next five years (1975–1980). Given that the total capacity of the entire nuclear supply industry in the United States was about 30 gigawatts per year, the forecast indicated that the supplier industry would operate at full capacity for some time—very happy news for GE's Nuclear Division. And the forecast was quite compatible with others made at that time: a popular number was that we would have 400 gigawatts of nuclear plants in the United States by the year 2000.

GE's chief planner was not comfortable, however, with the assumptions on the capital cost for the base case, and asked for a recalculation using 15% higher nuclear plant capital costs. (See Table 5-4.) Given this scenario, the model predicted that nuclear additions would fall off by a startling 40%. With the backlog already in place, the nuclear supplier industry would operate at 30% capacity for new orders—*exceedingly* bad news for GE's Nuclear Division.

**TABLE 5-3 Optimized Generation Planning: New Plant Additions, 1975–1990**

| *Plant Type* | *Backlog through 1985, in GWs* | *Base Case* |
|---|---|---|
| Nuclear | 90 | 242 |
| Fossil | 104 | 156 |
| Gas Turbine | 9 | 104 |
| Combined Cycle | 11 | 28 |

*Source:* Thomas H. Lee. 1975. Unpublished analysis, General Electric Co., Fairfield, CT.

**TABLE 5-4 Optimized Generation Planning: New Plant Additions, 1975–1990 (Raising Nuclear Plant Costs 15% from Base Case)**

| *Plant Type* | *Additions* |
|---|---|
| Nuclear | 138 |
| Fossil | 248 |
| Gas Turbine | 117 |
| Combined Cycle | 16 |
| Hydro | 28 |
| Total | 547 |

*Source:* Thomas H. Lee. 1975. Unpublished analysis, General Electric Co., Fairfield, CT.

At the same time, because the planner was worried about the rapid escalation of the construction cost for nuclear power plants and whether the utility industry as a whole would be able to raise the needed capital, a financial model was formulated for the entire utility industry in the United States. The model, applied to the base case, found that on an annual basis the industry might have to raise $30 billion per year, a reasonable challenge. But, what if construction costs escalated at twice the rate of inflation? The answer was that $80 billion per year would have to be raised—an impossible task.

What did this optimization exercise show? It produced a base case that forecast a healthy industry for suppliers and utilities, and a sensitivity analysis that indicated that a slight change in one assumption resulted in a drastic change in the outcome: a sick supplier industry and possibly a number of bankrupt utilities. The alternatives? One was to ignore the sensitivity analysis and stay with the base case. After all, the base case assumptions were the best guesses of the best experts. Hard as it may be to believe today, single forecasts were believed in many organizations. But GE's planners took one more step: they initiated a study to dissect the two most important assumptions in capital cost: economy of scale and capacity factor. We have already discussed these two wrong assumptions in Chapter 4. Now let's examine how one can use the results of that study in planning.

From the detailed cost study, the best one can conclude is that the strength of economy of scale ($A$) in the classic equation

$$\text{Cost} = K\frac{R^{-A}}{1000} \qquad \text{Equation 5-1}$$ [10]

lies somewhere between 0.3 and 0.5. Let's see what this equation means. $K$ is a constant and $R$ is the rating of the plant in 1,000s of megawatts. For example, if $R$ is 2,000 megawatts and if $A$ is 0.5, then the unit cost of the 2,000-megawatt plant is $K/\sqrt{2}$. Previously we have also shown that the decline in capacity factor ($C$) for every increase in the plant rating by 1,000 megawatts is probably between 0.2 and 0.4. For every pair of these two numbers ($A$ and $C$), there is an optimum size. (See Table 5-5.) For the ranges chosen, the optimum size varies from 395 megawatts to 1,300 megawatts, hardly useful for a planning decision.

If one plots, however, the entire curve of effective cost versus size, a different story emerges. We plotted three such curves for $A = 0.5$. (See Figure 5-1.) The optimum size for the three capacity factor slopes are indicated by arrows on the graph. If the capacity factor slope is 0.2, the curve in the neighborhood of the optimum is very flat. An error on the low side by 400 megawatts will only cause an increase in effective cost by 4%. On the other hand, if the capacity factor slope is 0.4, an error on the large side can significantly increase the effective cost. If the capacity factor is 0.2, in the range of interest (500–1,300 megawatts), the value of $A$ hardly matters; that is, *the economy of scale has no effect at all on the total cost.* (See Figure 5-2.) When the capacity factor is only 0.2, one can afford to choose the wrong size with very little penalty. For a slope of 0.4, it appears to be also true that $A$ has practically no effect. But an error from the optimum can cause significant penalties. Of course, for the same size,

**TABLE 5-5 Unit Rating for Lowest Effective Capital Cost**

| | | *Capacity Factor Slope (%/1,000 Mwe)* | | |
|---|---|---|---|---|
| | | 20 | 30 | 40 |
| **Effective Economy of Scale ($A_{eff}$)** | .3 | 1,090 | 730 | 550 |
| | .4 | 1,360 | 905 | 680 |
| | .5 | 1,580 | 1,055 | 790 |

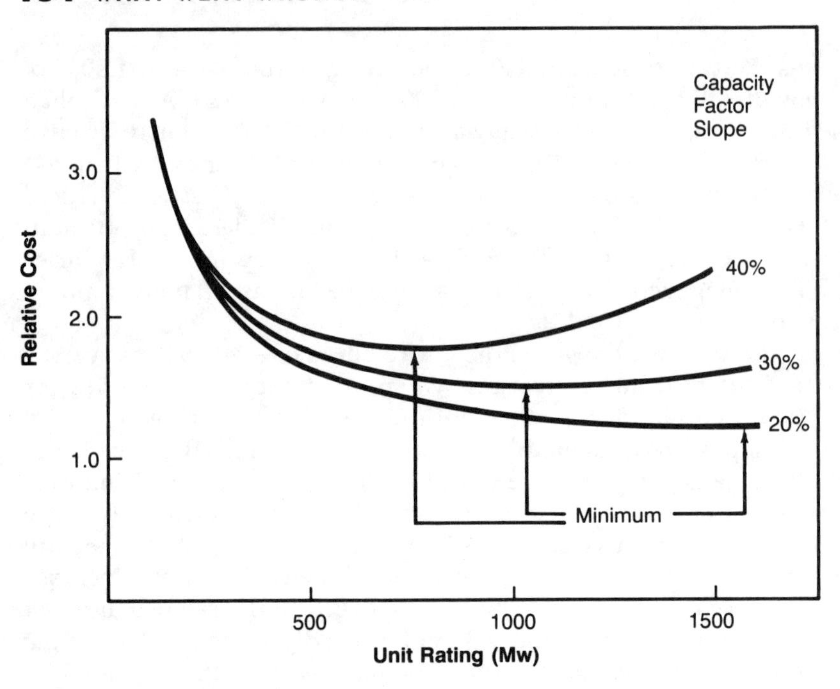


**FIGURE 5-1 Relative Costs of Power Plants for $A = 0.5$.**

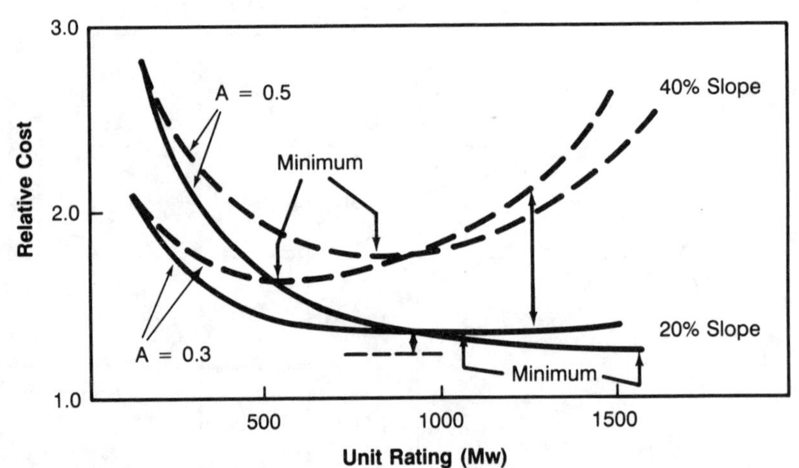


**FIGURE 5-2 Relative Costs of Power Plants for Two Sets of $A$ and $C$.**

the effective cost is significantly higher if the capacity factor slope is greater.

What does a low value of *C* mean? It means that the plant was designed with adequate margin so that scaling-up will not significantly affect its performance. This means a high-quality plant. Therefore, a quality strategy for this product is a much safer strategy than economy of scale, contrary to the conventional wisdom.

Several conclusions can be drawn from this case. The analysis on these two assumptions was made in the hope of finding the optimum size. But the quality of information was so poor that the optimum size was never found! Nevertheless, a valid strategy did evolve, even though the results from the optimization exercise were of very limited interest. The model was useful in identifying the critical parameters. Therefore, the most important challenge is to learn how to use the models.

### Limitations of Large Models

When one of the authors became the chief planner of GE's power generation business, the firm had already lost hundreds of millions of dollars in the nuclear business. It had just offered a new generation of boiling-water reactors (BWR-6, sixth design version), and had an order backlog, in total, for about sixty plants. At stake were several billion dollars. The challenge was to create a new strategy for the business.

In tune with the times, a young professor was hired to develop a large strategic planning model. The model performed quite well. It was capable of predicting the impact of postponements and cancellations of nuclear plant orders, second-guessing the long-range forecast of GE's Nuclear Division, and analyzing and comparing different strategic alternatives. The planner described the model to the group executive in charge of the business, who showed considerable interest and made some complimentary remarks. But the planner was never able to persuade the group executive to make a single decision based on the model. Looking back, we know this was a typical high-level executive reaction, not an isolated example. The executive simply did not trust the model because he did not know what was in it. The recent trends toward interactive systems indicate that the OR community is learning how to make its work more useful. But, in this chapter, we are talking about what went wrong. It is also important to point out that not all OR scientists agree with our view—a normal situation in the evolution of any science and technology.

A completely different approach at GE proved to be successful. The nuclear strategic issues were divided into a number of independent problems, for example,

- the question of optimum size (economy or diseconomy of scale)
- the reliability of a given plant design
- the terms and conditions of nuclear contracts
- regulatory practices and standardization

A team was assigned to work on each issue. Extensive analyses were made. Periodically, the entire management team would meet with and listen to study team progress reports. After two and a half years, the managers and analysts shared an understanding of the issues and each other's language describing the issues. It was not too difficult then to arrive at a viable strategy, which worked for many years.

What can we conclude from this example?

- It is difficult for decision makers to have confidence in large models whose inner workings they do not fully understand.
- It is important to create effective man-model interfaces by decomposing the problem.

Over the years, the International Institute for Applied Systems Analysis (IIASA) has developed many large models for global issues such as energy, food and agriculture, and forestry. When one of the authors became its director, there was an internal debate on the usefulness of the IIASA energy model. He decided to test the question on another model: the global trade model (GTM) for the forestry industry. The methodology used is traditional and the model is big, solid, good, and complex—but it is not user friendly in the sense that it could not easily be put to work by nonmodeler decision makers in the forest industry.

A model of this type has three potential uses: scientific advancement, as a tool for analysis and for predictions on the macro level, and as a tool supporting strategic planning and decision making on the corporate level. The GTM served the first two functions quite well. Whether the third use could be realized was not certain. An experiment was required.

M. Jouhki, an experienced forestry industry leader, president of Thomesto Oy, Finland, was invited to find out what aspects of the GTM could be used as part of the decision-making process in his firm. The goal was to interact directly with chief executives of selected forest companies in sessions simulating decision-making processes. The willingness of top managers to accept results from model runs and to take risks by modifying their positions based on information from the model was to be tested.

At the time of this writing, the experiment was almost completed. Four different tasks were planned:

- Analyze the evolution of forest industry management strategies during this century and describe management styles and attitudes, including its willingness to use analytical models.
- Create a set of guidelines for modelers on how to design a user friendly model. (This task is included because preliminary results already indicated the need.)
- Modify the user interface of the GTM to make it more readily accessible to nonmodelers.
- Evaluate results of planning sessions using the revised interface in a selected group of forest industry companies.

The preliminary conclusions are encouraging. For one thing, they show that the GTM can be useful without changing its comprehensive internal structure, but it should be made more user friendly. It was also evident that earlier interaction with users would have been wise.

The following is a list of the preliminary recommendations presented by Jouhki to the council of IIASA (Lee, 1989, pp. 2–3):

- Assumptions and boundary conditions are more relevant than analytical quality.
- Quantitative analysis alone leads to underutilization.
- To do right things is more important than doing things right.
- Let users ask the questions—early!
- Identify a small set of key issues.
- Modify systems for specific users.
- Constantly check assumptions.
- Not everything can be covered by mathematics.
- Test your findings continuously.
- The world needs simple models based on sophisticated assumptions—not the more popular sophisticated models based on simplistic assumptions.

This is the first time such an experiment has been conducted to the knowledge of the authors. When completed, we should know much more about the interaction between systems analysis and decision makers.

## Overreliance on Forecasts

An important application of models is forecasting—for demand, supply, population and economic growth, employment, inflation, and so forth. Our discussion of models would be incomplete if we did not address this application.

For the past five years IIASA and Stanford University have conducted surveys of long-term forecasts of energy demand, supply, and prices by well-known forecasters in many countries and held an annual International Energy Workshop to review the forecasts. We estimated prices of crude oil for the next two decades (see Figures 5-3 and 5-4). According to our estimates,

- The forecasts shift up and down, depending on the price of oil at the time of the forecast. (See Figure 5-3.) There appears to be no escape from the influence of the present.
- The variance from the mean for price forecasts is very wide, too wide to be useful for business planners. (See Figure 5-4.)

A similar situation can be seen in the forecasts published by the U.S. government. The U.S. Department of Energy's Energy Information Agency (EIA) prepared several forecasts of the price of oil.

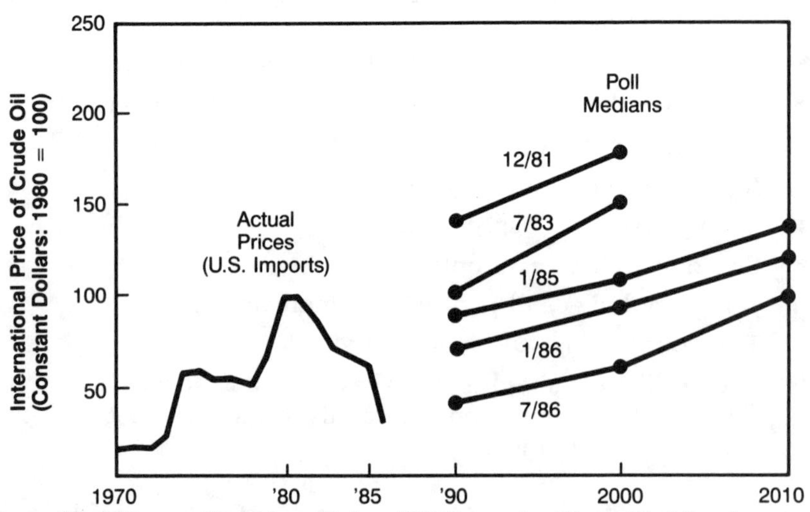


*Source:* Alan S. Manne and Leo Schrattenholzer. 1986. "International Energy Workshop: Summary of Poll Responses," IIASA. Laxenburg, Austria, January.

**FIGURE 5-3 Comparison of Five Successive IEW Polls and Actual Prices.**

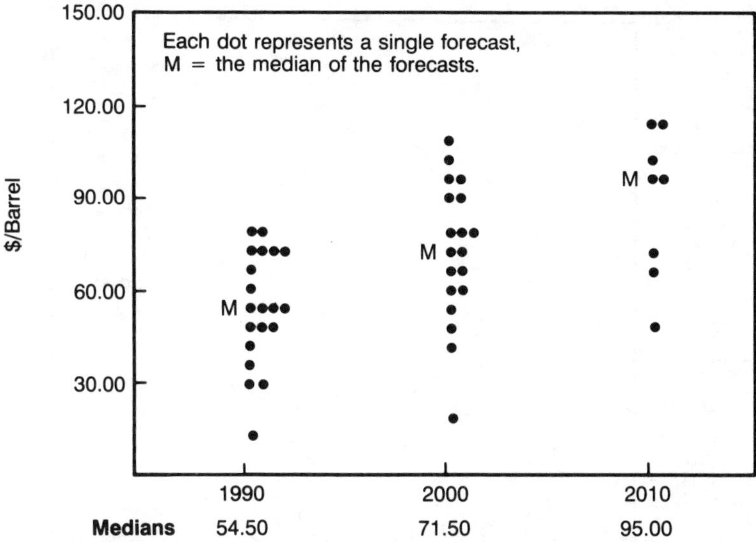


*Source:* Alan S. Manne and Leo Schrattenholzer. 1986. "International Energy Workshop: Summary of Poll Responses," IIASA. Laxenburg, Austria, January.

**FIGURE 5-4 1986 Forecasted International Price of Crude Oil.**

(See Figures 5-5 and 5-6.) Whatever today's price may be, forecasted future prices will be higher. Until we are able to forecast the kind of behavior that more closely parallels reality, midrange oil price forecasts will, in our opinion, continue to be misleading. (See Figure 5-7.)

How planners use the forecasts is a challenging issue. Their frustrations are undoubtedly what led some to say, "All energy forecasts are wrong." Kenneth Boulding reminds us:

> Computerized and numerical models, especially those with fancy diagrams and print-outs, are almost certain to produce illusions of certainty and many therefore easily lead to bad decisions. A study of computer-induced disasters, from bankruptcies to wars, is much overdue; we do not seem to have techniques for understanding uncertainty in the context of computerized models. (1974.)

## MISUNDERSTANDING THE DYNAMICS OF TECHNOLOGY

The dynamics of technology have been largely unappreciated by many policymakers. Two quotes popular during the 1970s serve to illustrate: First, President Carter's statement that the energy crisis

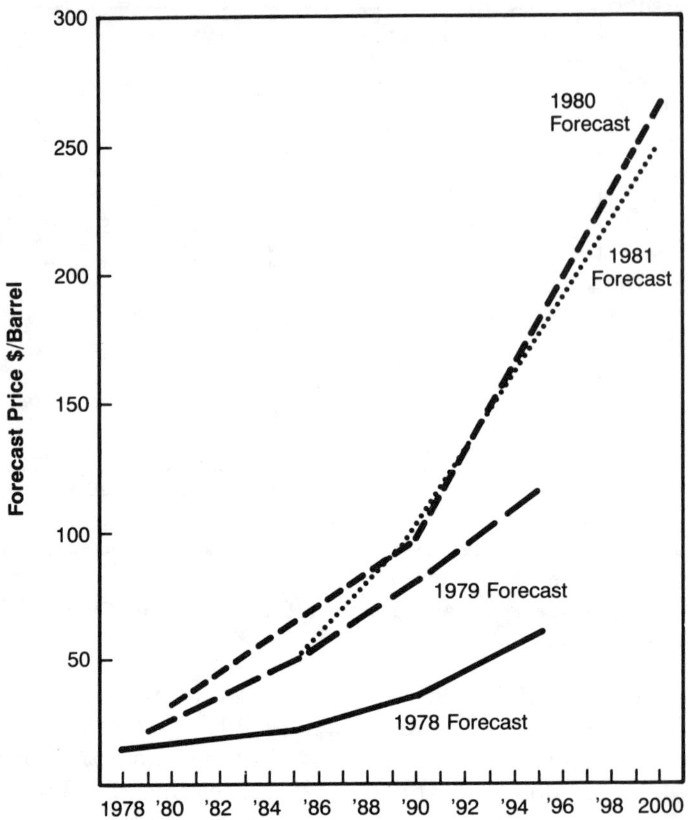


*Source:* Energy Information Agency, U.S. Department of Energy, Reports to Congress, 1978–1981, Washington, DC.

**FIGURE 5-5 Forecast of World Crude Oil Prices, 1978–1981 (current dollars).**

was "the moral equivalent of war." Second, the popular argument that "If we can put a man on the moon, why can't we solve the energy crisis?" The seductive simplicity and intuitive appeal of these quotes suggest why blunders about technology were made in the past and why they will be repeated unless we develop a more realistic appraisal of technology.

In this section we describe the process of commercial change, and how our misunderstandings led to sizable blunders, some still hidden from general view. One important point to be brought out is the limited effectiveness of government intervention in the dynamic. Then, we will discuss the concept of technological life cycle, which can be very useful in planning.

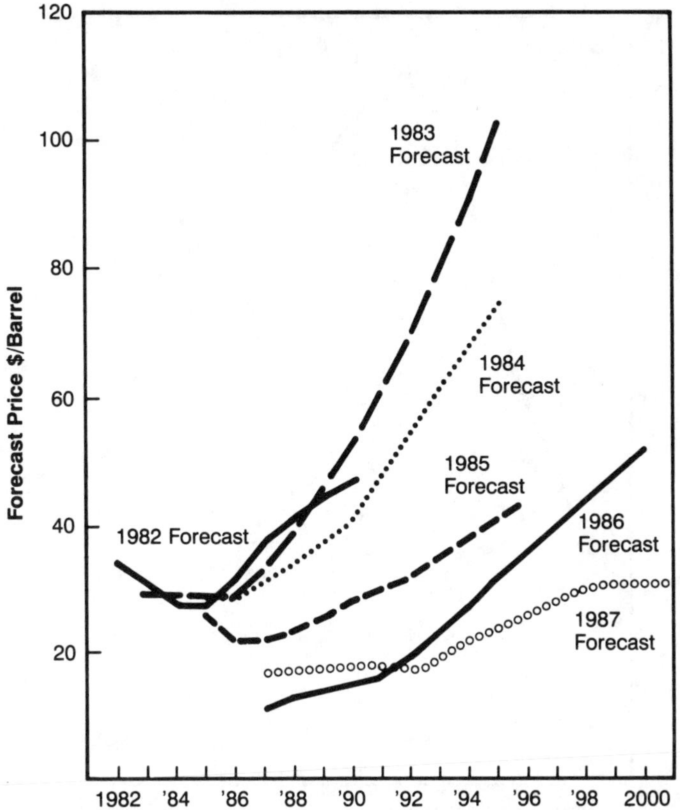


*Source:* Energy Information Agency, U.S. Department of Energy, Reports to Congress, 1982–1987, Washington, DC.

**FIGURE 5-6 Forecast of World Crude Oil Prices, 1982–1987 (current dollars).**

## Energy and Technology

Energy, in the popular conception, is inexorably bound to technology: not only the technologies of energy extraction (mining coal, petroleum exploration and production) but also the technologies used to convert raw energy into a useful form (refining crude petroleum into gasoline or diesel fuel, and burning coal, oil, or natural gas to produce electricity). These conversion processes are important both because they are large industries in absolute terms and also because they consume significant amounts of energy themselves.[11] And of even greater importance is technology's crucial role in the ultimate end use of energy: converting gasoline into private transportation, diesel fuel into public transportation, coal (with ore) into steel, gas into

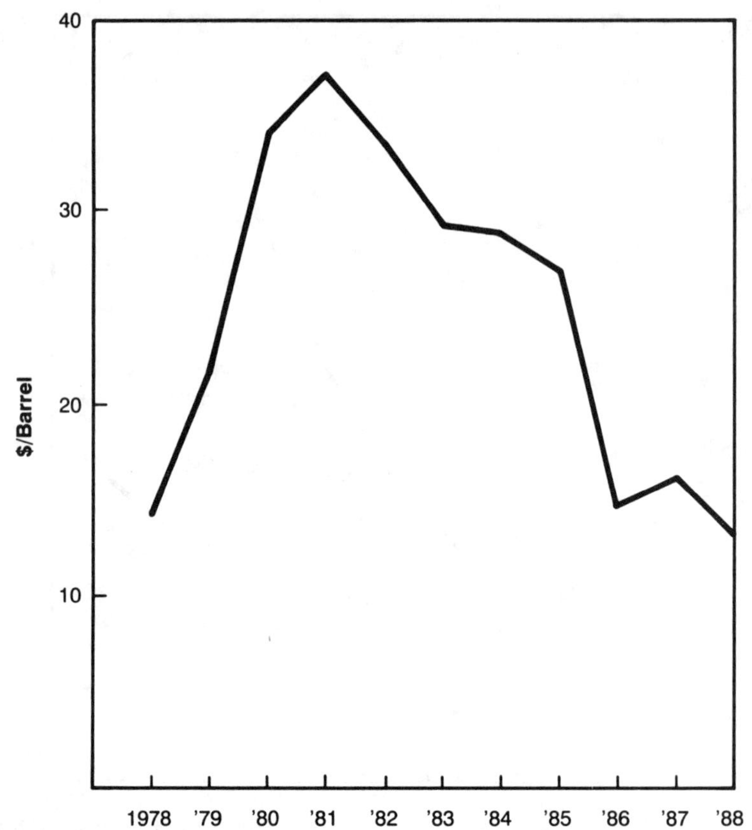


*Source:* Energy Information Agency, U.S. Department of Energy, Reports to Congress, 1982–1988, Washington, DC.

**FIGURE 5-7 World Crude Oil Prices, 1978–1988.**

heated buildings, and electricity into light. No one wants energy for its own sake. For example, gasoline has no value except as a fuel for vehicles. The automobile and engine technology are bound together in turning gasoline into its ultimate product: personal transportation. Furthermore, there are a host of technologies related to the automobile and its engine, such as steel making, autobody fabrication, welding, and so forth, all of which use energy.

The intrinsic relationship between energy on the one hand and the commonplace products and technologies we use on the other is so intimate and deeply embedded in our lives that we take it for granted: we rely on it every moment of our lives, yet we are aware of it only when it breaks down. It is no wonder that when the energy crisis appeared, we gasped and then resorted to technology for the solution. Unfortunately, in the process, we often neglected the dynamics of technology.

### Commercial Technological Change

Successful commercialization of a new technology occurs when it becomes available at a cost that allows the private sector to realize an acceptable return on the total capital required to produce and market it (Ball, 1981).

This statement seems true for those who function within a market economy. But three points need to be made. First, the discussion is not restricted to a perfect or ideal market, but applies to the market as it exists in reality—imperfections, government interference, and all.

Second, the discussion should apply equally well to planned economies. Although the lexicon, the arithmetic, and the quantities measured are different, the dynamics are the same.

And third, whether the economy is market oriented or planned, it is the consuming public that passes final judgment on the value of the technology. Whether the public votes with its dollars or whether it expresses its preferences and values another way, public choices and values ultimately are the determining factor in the commercial success of any technology. Therefore, when we say that a technology is "best" or the "most cost effective," we do not mean it in an abstract, technocratic, or theoretical sense. We mean "best" as determined by the relevant marketplace. Ours is not to judge; the marketplace does that. The challenge is to replace inappropriate contexts, criteria, measurement systems, accountability systems, conceptual frameworks, and analytical tools with the most appropriate ones possible. Only then will the resulting decisions and judgments be the most viable and appropriate possible.

How, then, does a technology reach the point of public acceptance—of commercialization? The following discussion answers the question, and shows how mistaken assumptions lead us into grievous errors about energy technologies.

The successful commercialization of a new product goes through five discrete stages. These stages apply equally to energy-related technologies.

- Invention
- Development
- Commercial introduction
- Commercial diffusion
- Established stage

The first stage—invention (or, in some cases, research)—begins with the generation of an idea for a product, along with the conceptualization of a way to produce it. At this stage, because technical, market, and regulatory uncertainties are very high, and because costs, prices, and markets usually are poorly known, economic considerations are not a factor.

In the development stage, the design of the product is optimized until it is possible to construct a working model. The principal function of this stage is to eliminate such factors as technological uncertainties and to improve operational efficiencies, with an eye to determining expected costs and variability of eventual production. The development stage need not, and usually does not, require the construction of a full-scale production facility. Much more frequently, the approach is to build a single pilot plant. (This could be a scaled-down version of a machine or piece of equipment; for the sake of simplicity, we will use the word *plant* throughout.) The *product* may be real—a cubic foot of natural gas, shaft horsepower, BTU, or a Cabbage Patch Doll—but usually the pilot plant is one or more orders of magnitude smaller than full size. When the important factors of production have been realistically scaled, the pilot plant is used to yield information as to whether there are significant technical uncertainties looming ahead, and whether these uncertainties must be resolved by constructing a full-scale production facility. But if a small subunit of the plant is not readily scalable, then it may become necessary to build that subunit large enough to resolve its technical uncertainties. Only in the rarest of instances is an entire full-scale production facility required at this stage.

Thus, in addition to the reduction of technological uncertainty, the purpose of the development stage is to determine the expected costs of production in the established stage.

At the end of this stage, variance from the established costs will be high; at this point, however, the expected value of the costs has been determined. If the costs are too high to offer hope of adequate profit, then the technology is not viable commercially. In this event, the project is dropped or, in some cases, it is put on hold in the hope that a new factor will some day reduce costs.

The development stage is not primarily intended to reduce costs, but rather to determine them, and like the previous stage, it does not deal with market or regulatory uncertainties.

The third stage—commercial introduction—does address these issues. At the same time it further narrows the range of cost variance. At this point, full-scale production facilities are put into operation and a number of market uncertainties are rigorously studied: marketing programs, distribution channels, maintenance support organizations, market segmentation and differentiation, character of the technology and the industry, and so forth. Regulatory and legislative uncertainties (e.g., environmental, tax, and procedural delays) are explored.

Only after the cost, market, and regulatory uncertainties have been addressed does the fourth stage—commercial diffusion—begin. This stage is marked by widespread production in a growing number of full-scale facilities, widespread dissemination into the marketplace, and the likely entry of market competitors with similar products. Costs begin to decrease because of product and process innovation.

Finally, in the established stage, as experience is gained, costs continue to decrease, even though by now the product (or technology) is maturing, as are production processes and basic marketing strategies. Most activity focuses on fine-tuning marketing strategies. The continuing decrease in cost in this stage is a well-known phenomenon in manufacturing processes known as the learning curve, although it also applies to the other functions: management, distribution, marketing, and so forth (the experience curve) (Allan, 1975). This cost reduction is a function of the number of *comparable* units produced, and is expected to account for a 20% to 30% cost reduction with each doubling of accumulated experience. On a log-log plot, the curve (unit cost versus accumulated experience) is a straight line with a slope of 70% to 80%. The early points on the plot are scattered and do not begin to take form until the system as a whole becomes organized and routine, which happens only well into the commercial diffusion stage.

At first glance, the five-stage process seems simple and straightforward. One discrete step follows the next in a logical and ordered fashion. The commercialization of technology has parallels to human growth and development: we learn to crawl before we learn to walk; and we learn to walk before we learn to run. This is not a process to be toyed with. Accelerating one stage inappropriately or fusing two stages in haste to move a product or technology forward can result in disaster. This is precisely what happened with several energy technologies in the 1970s.

After the OPEC oil embargo and subsequent price spikes in the early 1970s, one of several major responses by the U.S. government was to encourage the exploitation of alternative, domestically produced energy: synthetic fuels, photovoltaics, wind-powered generators, and so forth. The approach was to build *commercial-size* plants, with government support and money, in order to demonstrate that such new energy technologies were commercially feasible. *Commercial demonstration* was the order of the day.

The assumption behind the approach was that the private sector, having once seen full-size plants actually operating, would replicate them on its own, thus bringing into being new energy technologies much more quickly than would have been possible otherwise (Ball et al., 1976). The entire nation would benefit from the resulting reduction in U.S. dependency on imported oil.

One effort involved the planning, development, and construction of several synfuels plants, which would derive various liquid and gaseous synthetic fuels from coal or oil shale. The problem underlying this approach was that the dynamics of technology were not well understood, either by industry or by government policymakers (Ball and Hammond, 1977). In the five-step process described above, it is the construction of small-scale pilot plants in the development stage that proves the viability of a new technology. Admittedly, there are questions that pilot plant operation cannot answer, reliability being one of them. Most of these should be addressed by careful analysis such as failure mode and effect analysis, and also extensive component testing. To answer these questions with a full-scale operating plant is a needlessly expensive approach. We did it in the nuclear business, and the low capacity factor of nuclear power plants is a consequence. Unfortunately, in the synfuels case industry and government proceeded to do exactly this: build full-scale synthetic fuel plants at a cost of billions of dollars. The result has been at best very disappointing. Those still in operation have extremely high product costs compared to those that were anticipated. In fact, full-size plants using the synfuels technologies in which the U.S. government was interested had been in operation for some time (e.g., shale oil in South Africa and coal liquefaction in wartime Germany). The plants were technically feasible, but not economically so.

Government and industry ignored these facts, just as they ignored the five-stage model of successful commercialization. They built the demonstration plants as if they had progressed to the commercial introduction stage—as if the economics were proven. In reality, the economics were known to be noncompetitive, and the technology had already been proven.

The demonstration approach by government and industry ignored a critical point no successful entrepreneur could afford to overlook: entering the commercial introduction stage makes no sense unless the technology earlier has been shown to be cost competitive *in the development stage.* And synfuels simply were not cost competitive, not by a long shot. The government's approach to synfuels assumed that the private sector had not entered the introduction stage on its own because of technological uncertainties. So if it overcame them, synfuels would become widespread. But the simple truth is that the private sector had not entered synfuels production because it *already* had determined that expected costs were too high to permit adequate returns. And since the government demonstration program did nothing to change the *costs*,[12] the effort was doomed from the start. In fact, as our model of technology dynamics indicates, if the expected cost of synfuels had been cost competitive, then the private sector would have entered the introduction stage, that is, built demonstration plants, speedily and on its own, regardless of government encouragement and support. If the expected cost of shale oil, based on results gleaned during the development stage, was $8 instead of $80 per barrel, industry would have charged ahead without government support. No wonder there are no new synfuels projects.

After the embargo of 1973, demonstration-mania swept the United States, spreading to many new energy technologies. Some failed because they were not cost competitive—solar photovoltaics and wind-generated electricity for input to electric utility grids are two examples. Others were already cost competitive, and did not need government support to begin with. They included increasing the fuel efficiency of motor vehicles, increasing the efficiency of heat recovery in combustors, and adding insulation and improved temperature controls in commercial and residential structures.

We should add that perhaps we have learned our lesson, since in the "clean coal" program—a particular example of a synfuel—more emphasis is placed on research than on demonstration plants.

In summary, the experience of the 1970s and 1980s taught us that *if a technology is commercially viable, then government support is not needed; and if a technology is not commercially viable, no amount of government support will make it so.*

**Bad assumptions.** The synfuels demonstration program was based on two additional unsound assumptions. The first was that world oil prices would continue escalating without ceasing. This assumption suggested that if a new technology were not cost competitive today, it would become so at a future time. But time proved otherwise. It was the expected costs of the new technologies that continued rising, and, beginning in 1981, it was the price of world oil that began to fall. Ironically, oil was taking its deepest price dives at about the same time really big money was being spent on the construction of synfuels demonstration plants.

The second unsound assumption was based on another generic misunderstanding of the dynamics of technology, a misunderstanding about the significance of the experience curve. The assumption was that costs would decline as a developing technology progressed along the experience curve; that is, a second commercial plant would cost 75% of the first, the next two would cost 75% of the second, the following four would cost 75% of the previous two, and so on. The assumption was that unit costs would be 25% lower with every doubling of production.

The error in applying this principle to demonstration projects was overlooking the effects of *shared* experience on costs. For example, a wind-powered electric generator is built with components that have a long history: steel towers, propellers, gear boxes, generators, and transformers. All are well-developed and well-understood components of other technologies. Therefore, building a second windmill does not *double* the number of experiences on the experience curve; at best, it merely adds an additional unit of experience to an already well-understood technology. In fact, doubling the accumulated experience of each component of a windmill would take a long time.

In point of fact, when experience-curve-related cost reductions do exist, they apply primarily to technologies in which major components are truly *new*, for example, most of the components of jet engines when they were introduced twenty-five years ago. Again, only those costs *unique* to wind-powered electric generators could reap significant benefits of the experience curve (Ball, 1979). In most new energy technologies of the 1970s and 1980s, the crucial components already had exploited the experience curve. Most of the benefits had been gleaned. Thus, many of the cost components of shale oil plants shared experience with other kinds of materials handling and processes, plant facilities, and so forth.

An additional misunderstanding about the experience-curve effect is the assumption that cost reductions would begin to accrue at the moment the first demonstration plant went into operation. In fact, cost reductions do not begin until the system *as a whole* is organized and operating routinely, which only happens well into the commercial diffusion stage. The basic premise underlying the experience curve simply is that doing the same thing becomes easier, and hence cheaper. Prior to the commercial diffusion stage, because the tasks are not routinized, experience-curve cost reductions cannot be realized.

The U.S. nuclear industry provides an excellent example of shattered hopes that were placed in experience-curve cost reductions. Such reductions depend on *learning* through *repetition*. The repetition simply did not occur in the design and construction of U.S. nuclear reactors. In the United States, practically every nuclear plant was custom designed and custom built. No two were alike. A consequence of such a policy is that no experience-curve cost reductions are possible, nor can they reasonably be expected.

By contrast, the Canadian Candu[13] nuclear power plant and the light-water reactors (LWRs) in France and Japan quickly became standardized in both design and construction. Despite the high initial capital cost of the Candu reactors, the experience-curve cost reductions that one would expect have actually been achieved, bringing the cost of the plants and the electricity they generate to satisfactory levels. France and Japan are experiencing similar reductions.

## A Life Cycle for Technologies

The five stages in technology change describe the process from the viewpoint of decision makers and the realities of the commercial mechanisms. There is a different way of describing the dynamic of technologies—the concept of technology life cycles. The idea of a life cycle goes back to the 1930s when a general "law of industrial growth" was proposed (Alderfer and Michl, 1942). Subsequently, many researchers studied the process of industrial development. They include Kuznets, Burns, Hoffman, Hansen, and more recently, Abernathy and Utterback (1978). Miller and Friesen (1984) compiled a great amount of empirical data. There is no doubt that regularities exist to confirm the validity of the life cycle concept. An excellent review of past research and publications can be found in Ayres (1987). Ayres clearly pointed out that in spite of the regularities, economic theory has not been able to explain the dynamic of innovation and technological change. But for decision makers and business people, the existence of regularities should be enough reason not to ignore the concept.

The basic concept is that technological systems are similar to biological systems. One can define in a qualitative way three stages in the life cycle: the embryonic, growth, and maturity stages. (See Figure 5-8.) We can compare the relationship between the five stages in the process of commercial technological change and the three stages in the technology life cycle. (See Figure 5-9.) Although there is some correspondence between the two models, they illuminate different issues.

Each stage has its unique characteristics. In the embryonic stage, an industry is usually characterized by large numbers of new ideas and large numbers of entrepreneurs competing for the same niche. Many entrants fail at this stage. The weaklings are eliminated; the hardy survive.

The survivors pass into the growth stage. The number of competitors is further reduced. Scale economies and experience curves raise barriers to new entrants. Large-scale production and productivity improvements permit cost reduction and lower prices. Nonzero price elasticity of demand translates into larger markets. During this stage, the industry may enjoy its maximum rate of expansion.

The third stage (maturity) is reached when markets are becoming saturated and the price elasticity of demand falls toward unity. In this stage, product innovation is apt to be slow or nonexistent. The

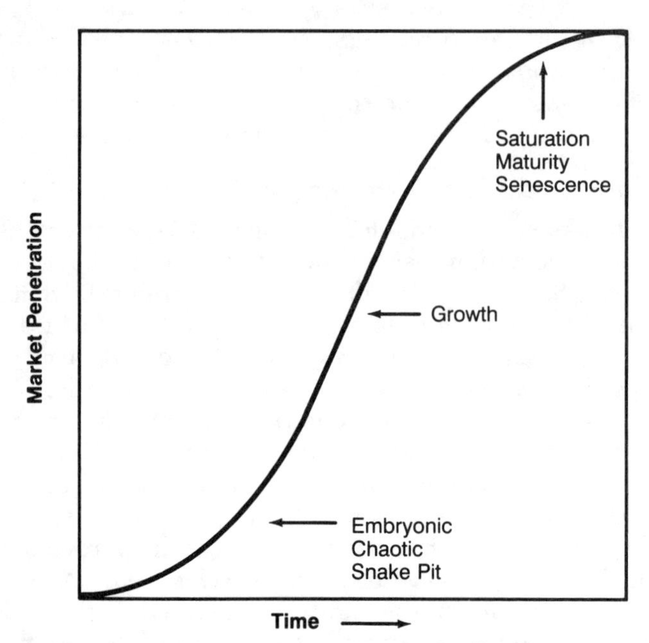


**FIGURE 5-8 Three Stages of the Life Cycle Concept.**

industry tends to be dominated by a few large firms and production capacity tends to move from high-wage to low-wage countries, as observed by Vernon (1966). In general, the industry is plagued by the serious problem of excess capacity.

What drives the shift from one stage to another? There are many driving forces: market growth, evolutionary advances of technologies, reduction in the cost and price of the products, and so forth. It is interesting to find that many, if not all, change with time in a way that is consistent with the life cycle concept; that is, they evolve along S-shaped curves.[14] This is what we mean by regularities. Before

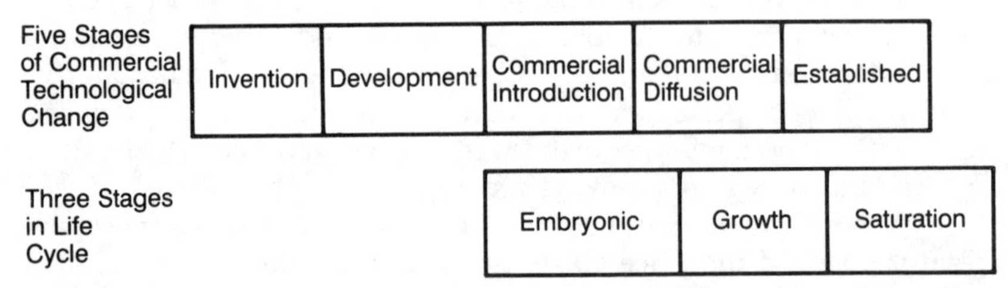


**FIGURE 5-9 Comparison of Five-Stage and Three-Stage Theories.**

we discuss how these regularities should be considered in business decisions, let us first review some observations in the energy sector.

The world commercial market for liquefied natural gas (LNG) follows the simple logistic equation well, with a "delta *t*"—the time it took to change from 10% to 90% of the ultimate level—of twelve years (Marchetti and Nakicenovic, 1988). (See Figure 5-10.) The necessary LNG tanker fleet for the market took ten years to develop, very close to the figure of twelve for the total market.

The growth of natural gas pipelines in the United States, a measure of the size of the gas market, also appears as an elegant S-curve (Grubler and Nakicenovic, 1988). (See Figure 5-11.) The maximum depth of exploratory drilling—a measure of technical progress as a function of time—follows a similar curve. The same effect is demonstrated by the efficiency of prime movers. That the different parameters follow the simple S-curve equation indicates that there are indeed regularities one must pay attention to.

The question is, How can these regularities be helpful in business planning and decisions? We will use two examples: one in the aviation industry, which is part of the transportation sector, a big user of energy, and one in the electric power industry.

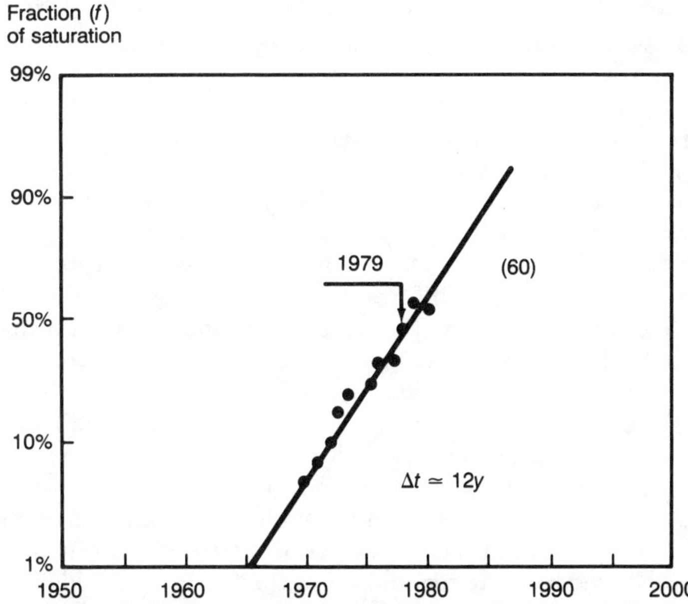


*Source:* C. Marchetti. 1988. "The Future of Natural Gas," with T. H. Lee, ed. *The Methane Age.* Dordrecht/Boston/London: Kluwer Academic Publishers.

**FIGURE 5-10 World Commercial Market for Liquefied Natural Gas.**

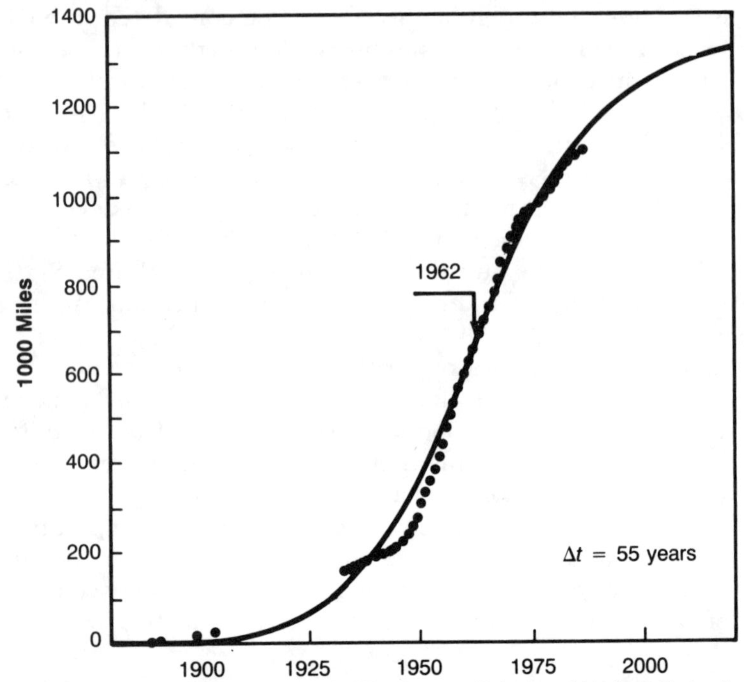


*Source:* A. Grubler and N. Nakicenovic. 1988. "The Dynamic Behavior of Methane Technology," in T. H. Lee, ed. *The Methane Age*. Dordrecht/Boston/London: Kluwer Academic Publishers.

**FIGURE 5-11 Growth of U.S. Natural Gas Pipelines (Length).**

The evolution of the technical performance of passenger aircraft appears as a band instead of closely following a straight line (Lee and Nakicenovic, 1988). (See Figure 5-12.) The left line represents the performance of the best airplanes. If one had to make a choice between different modes of transportation for investment purposes in the 1930s, the most favorable information available on aviation was for the DC-3. Comparing the performance, cost, and personal comfort of a DC-3 with that offered by railroads, one might easily conclude that the railroad would remain superior. Fifty years later, there is no way to travel between the coasts in the United States by rail. And the reason is that the young aviation technology in the 1930s has improved its performance (as measured by passenger km/h) by a factor of more than 100. Associated with the improvement in performance are cost and personal comfort. There was no comparable improvement in the already matured railroad technology of the 1930s. It is evident, therefore, that business decisions based on the economics of technologies in different parts of their life cycles will not be sound.

At this point readers might wonder whether there are internal

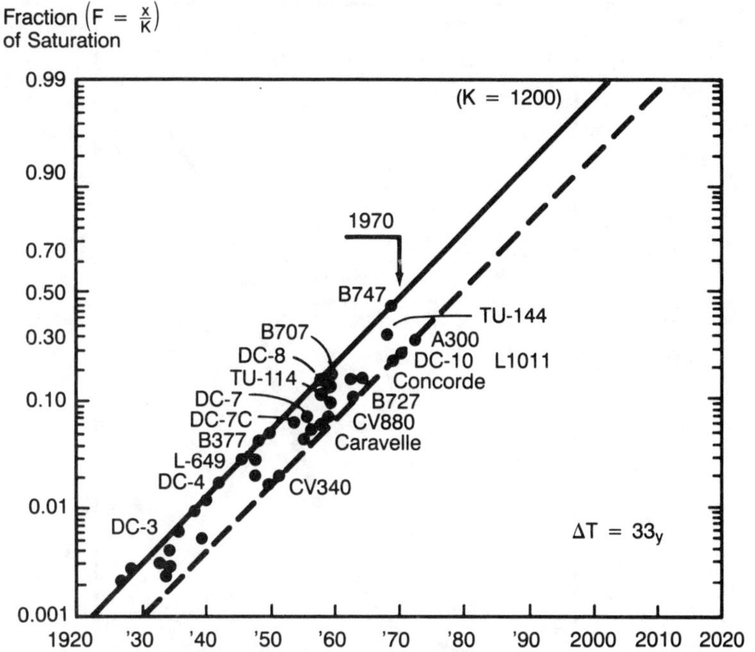


*Source:* T. H. Lee and N. Nakicenovic. 1988. "Technology Life Cycle and Business Decisions," *International Journal of Management,* Vol. 3, No. 4, p. 418.

**FIGURE 5-12 Improvement of Passenger Aircraft Productivity in Thousand Passenger-Km/h Plotting F/(1 − F) as a Function of Time on Semi-log Paper, Where F Is the Fraction of the Estimated Saturation Level K.**

inconsistencies between what we are saying and our discussion on synfuels. We don't believe so. Some developing technologies can overcome an early cost disadvantage by serving a niche where nonquantifiable advantages play a very important role. In the case of aviation, there was a market, though small, where cost difference between the new mode of transportation and the traditional one is less significant than the excitement in the increase of speed. In the case of synfuels, no such niche existed.

The world air transport market, measured in passenger Km/year $\times 10^6$, follows an S curve (Lee and Nakicenovic, 1988). (See Figures 5-13 and 5-14.) If the trend continues, the market will be saturated at $200 \times 10^6$ passenger *Km/y*, that is, $K = 200$ (Figure 5-14). The time, delta $t$, for the market to develop from 10% to 90% of the ultimate level is twenty-nine years. Although the implications are interesting, even more can be extracted if we combine the information (see Figures 5-12, 5-13, and 5-14):

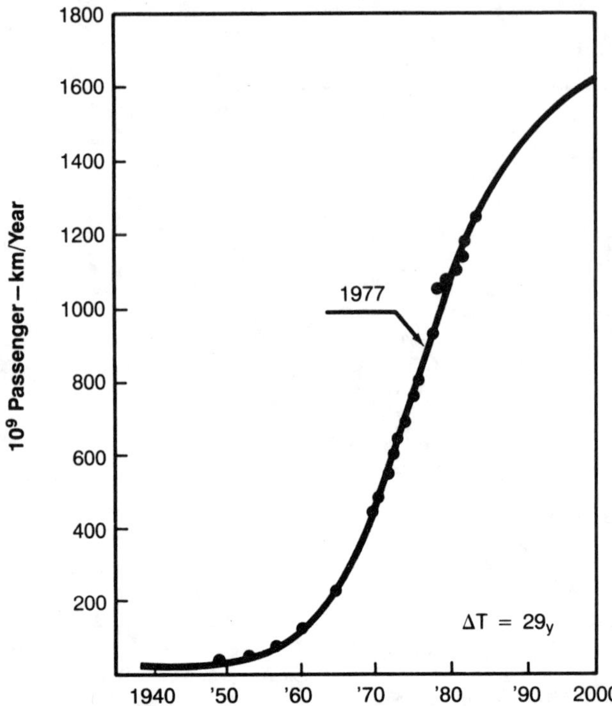


*Source:* T. H. Lee and N. Nakicenovic. 1988. "Technology Life Cycle and Business Decisions," *International Journal of Management*, Vol. 3, No. 4, p. 416.

**FIGURE 5-13 Growth of World Air Transport in Billion Passenger-Km per Year.**

- The performance of the present 747 is almost halfway to the ultimate level. A stretched 747 may be all that is needed. Certainly, this has serious implications on R&D decisions.
- The ultimate market can be served by 340 planes of the present vintage, if the aircrafts operate to capacity all the time. (Dividing the saturation value *K* in Figure 5-14 by the performance of the 747 in Figure 5-12.)

Of course, this is not so for many reasons. Nevertheless, there are about six hundred 747s in service now, suggesting approximately a 30% utilization factor.[15] The obvious conclusion is that the need for future production capacity may not be great. Yet at present, both Japan and the EEC are trying hard to increase their penetration of the market. Cutthroat competition and excess capacity may be inevitable.

It is also interesting to note that delta *t* for the evolution of technical performance (see Figure 5-12) is thirty-three years, not very

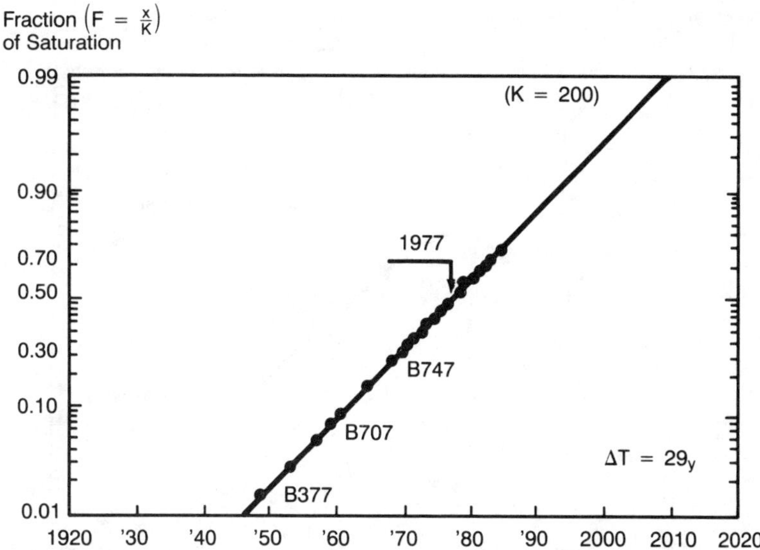


*Source:* T. H. Lee and N. Nakicenovic. 1988. "Technology Life Cycle and Business Decisions," *International Journal of Management*, Vol. 3, No. 4, p. 417.

**FIGURE 5-14 Growth of World Air Transport in Million Passenger-Km/h.**

This plot shows saturation fraction as a function of time on semi-log paper. In this way, results that would otherwise appear as S-shaped curves come out as straight lines, making them easier to interpret.

different from that of the total market. Associated with the dynamic technical evolution is cost improvement, which has a significant effect on market growth. At the embryonic stage, cost improvement is easy to achieve. The market grows rapidly, because of price elasticity. When a product moves toward the mature stage, improvement in technical performance becomes more difficult, the learning effect on cost slows down and so does market growth. It is therefore not surprising that the delta *t* for a purely technological attribute is not very different from that of the market attribute.

Blind belief in pure S curves can be dangerous, however, and the aviation industry provides us with evidence of that.

Nakicenovic (1987) studied the substitution of different technologies in the transportation market. (See Figure 5-15.) Thus in 1940, one would see a young growing aviation market, but to move forward aggressively with the dominant technology would have been a mistake. If one looks at the performance of the piston engine (see Figure 5-16), its performance at that time is saturating, since $F$ is greater than 0.9 by then. On the basis of these two graphs, one would be inclined to follow the market-share strategy, which suggests increas-

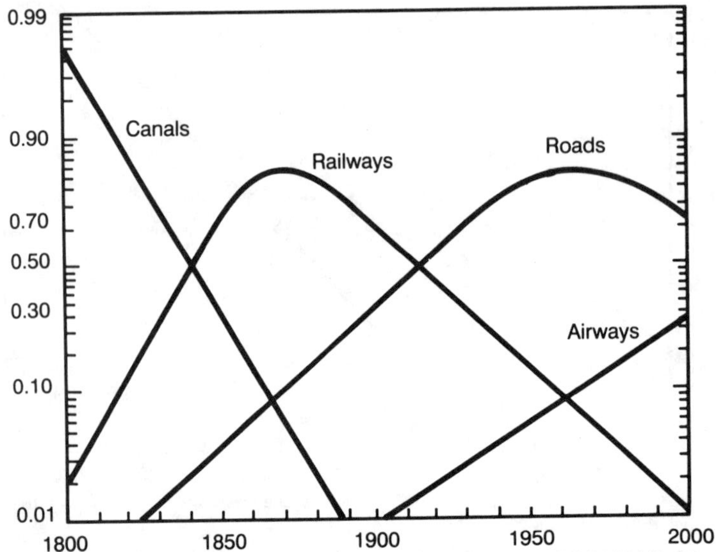


*Source:* N. Nakicenovic. 1987. "Transportation and Energy Systems in the U.S.," IIASA Working Paper Number WP-87-1, Laxenburg, Austria.

**FIGURE 5-15 The History of Transport Infra-structure Substitution in the U.S. from 1830 to 1982.**

Market shares of transport infrastructures are projected through 2000. F is the fractional share of a given transport infrastructure length in total length of all transport networks.

ing production capacity of piston engines. That would be disastrous, because a new technology—jet engines—soon appeared to displace the piston engines. The advances of capacity for the best engines on the market are two straight lines parallel to each other. This means that there are two S curves, one sitting almost on top of the other (see Figure 5-17), each describing the evolution of a technology, the bottom one for piston engines and the top one for jet engines. Delta *t* is the same for both technologies, indicating the existence of social, economic, or environmental forces common to both. In this case, they serve a common market, aviation.

The combination of an expanding market and a saturating technology may be the stage for the appearance of a new technology, a dynamic phenomenon that was not appreciated in the past.

An almost identical situation occurred in the electric power industry, with the evolution of high-voltage transmission systems

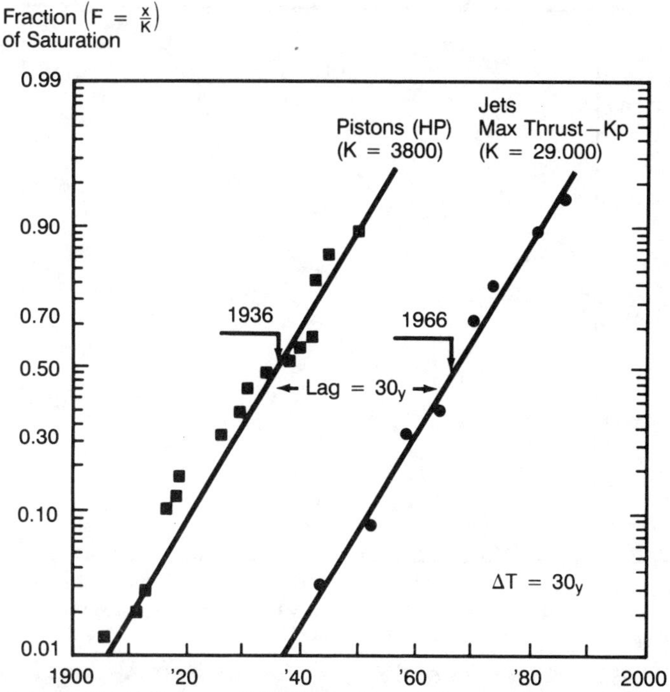


*Source:* T. H. Lee and N. Nakicenovic. 1988. "Technology Life Cycle and Business Decisions," *International Journal of Management*, Vol. 3, No. 4, p. 422.

**FIGURE 5-16 Advances of Capacity for the Best Aircraft Engines on the Market.**

The plot shows two S-shaped growth pulses. F is the fraction of the estimated saturation level K. The first pulse (left) shows piston aircraft engines and the second (right) jet engines.

(Lee and Nakicenovic, 1988). (See Figure 5-18.) In 1940, the left curve indicates that transmission voltage would saturate at about 300 kv. But if one examines the growth in electric demand at that time (Figure 5-19), it was still in the lower half of the S curve. Again, if one decided to follow the learning-curve strategy, the results might be disastrous. What actually happened? A second S curve (right curve in Figure 5-19) appeared. Utilities discovered that if they shifted their investment from generation to transmission, the overall economics improved. An international competition on who has the highest voltage system began. This pushes the delta *t* of the second S curve to nineteen years.

We can conclude, therefore, that looking at the S curves of a number of attributes can be useful for business planning.

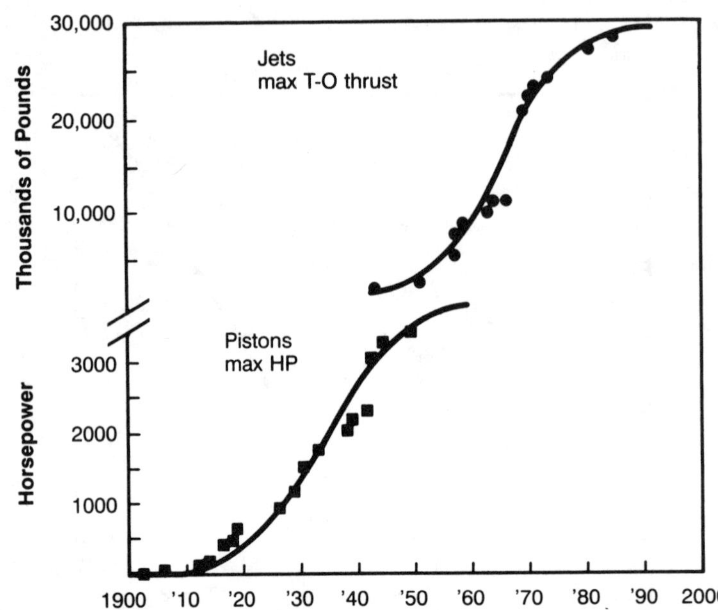


*Source:* T. H. Lee and N. Nakicenovic. 1988. "Technology Life Cycle and Business Decisions," *International Journal of Management*, Vol. 3, No. 4, p. 422.

**FIGURE 5-17 Advances of Capacity for the Best Aircraft Engines.** The first (lower) S-shaped growth pulse shows the increase in horsepower (hp) of piston aircraft engines, while the second (upper) growth pulse shows the increase of take-off thrust (in thousands of pounds) of jet engines.

A very important aspect of the life cycle concept deals with the competition between technologies. Almost twenty years ago Fisher and Pry (Fisher and Pry, 1971) demonstrated that when two technologies compete for market share, the process also follows the simple logistic equation; that is, the market shares of both winner and loser follow S curves. Marchetti extended their approach to the competition among energy sources. Let us now review that work.

**The dynamics of substitution.** Over the last 150 years, coal replaced wood and oil replaced coal; gas and nuclear are now replacing oil. (See Figure 5-20.)

Predictions made by C. Marchetti on energy substitution were controversial and not broadly accepted. Nevertheless, his analysis is fascinating, and does demonstrate in a dramatic way three historical facts:

- The time required for one energy technology to substitute for another is long—in the neighborhood of fifty-five years.

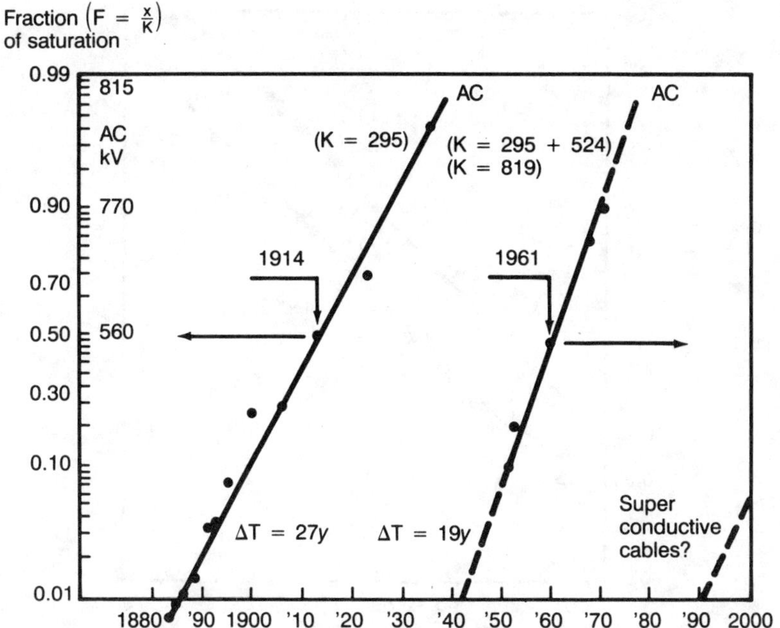


*Source:* T. H. Lee and N. Nakicenovic. 1988. "Technology Life Cycle and Business Decisions," *International Journal of Management*, Vol. 3, No. 4, p. 422.

**FIGURE 5-18 Evolution of High-Voltage Transmission Systems (kilovolts).**

- New energy technologies enter the market, gain market share, and finally fade away in remarkably consistent patterns.
- The major energy trajectories have never reversed themselves.

The fatalistic, definitive nature of much of Marchetti's work need not be debated here. Nevertheless, the time required to effect energy substitutions may be fundamental in nature and should not be overlooked—whether the context be policy, competition, or R&D.

One might not believe in the logistic predictions of Marchetti; namely, after oil, natural gas is destined to become the dominant energy source. But as prudent planners, policymakers, and business leaders, we at least should ask: Do we see young technologies today that might propel the position of natural gas into prominence? If one had asked such a question in 1975, the R&D priorities might have been different. For example, the gas-turbine technologies were in a younger stage than steam-turbine technologies:

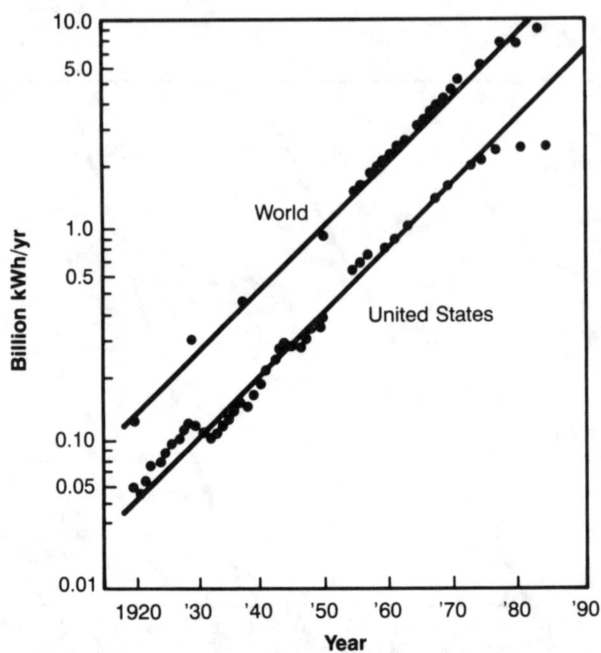


*Source:* T. H. Lee and N. Nakicenovic. 1988. "Technology Life Cycle and Business Decisions," *International Journal of Management*, Vol. 3, No. 4, p. 422.

**FIGURE 5-19 Growth in Electric Demand (Billion kWh/yr).**

- The firing temperatures and cooling technologies have been advancing rapidly; the corresponding situation in steam turbines showed all the signs of maturity.
- The combined cycles barely came into the picture. No attempt was made to optimize performance.

The associated technologies were in similar relative positions:

- Drilling technologies were advancing: deeper and faster. Meanwhile, coal-mining technology was maturing, and was severely hampered by institutional constraints (including labor) and by growing concerns over health and safety issues.
- Boilers and reheating cycles in coal-fired systems are old, whereas combustion technologies in gas turbines are young and reheating has not even been explored.

Our national policy decision not to consider natural gas was based completely on the running-out assumption. Technological

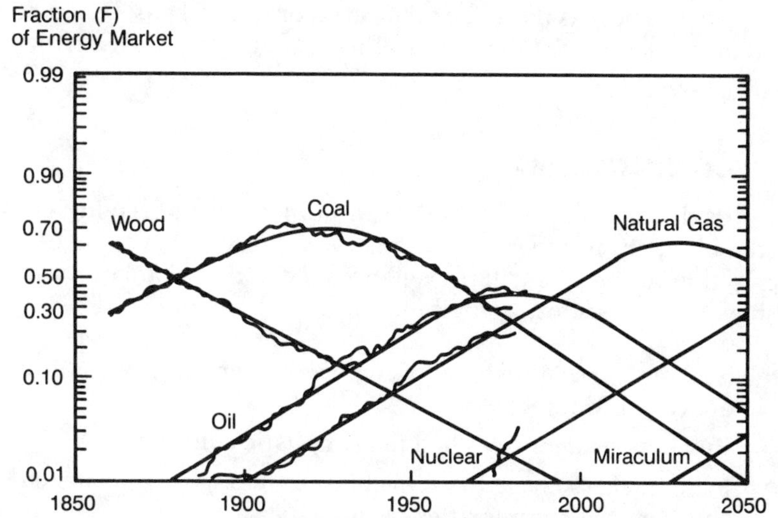


*Source:* C. Marchetti and N. Nakicenovic. 1984. "The Dynamics of Energy Systems and the Logistic Substiution Model," Research Report 79-13, IIASA, Laxenburg, Austria.

**FIGURE 5-20 Historical and Projected Trends in Global Primary Energy Consumption.**
The amount of energy (in coal BTU equivalent) from each source is plotted as fraction *f* of the total energy market. The smooth trends are the estimates based on historical data; scattered lines are historical data; straight lines show where energy sources follow logistic substitution paths.

dynamics were never an issue. We never considered that the available resource could be a function of technology.

Now, more than ten years later, we observe the prospective advances in gas technologies more clearly:

- Exploration for gas has shifted toward a search for gas per se. Previously, much of the gas was discovered as a by-product of the search for oil.
- Remote sensing by satellites and ground truth measurements are making inroads in exploration technologies.
- The efficiency of combined cycle systems is moving toward 52% or higher. Japan is starting its own gas-turbine industry.
- Active R&D projects are under way to make more effective use of natural gas, such as partial oxidation and the direct conversion of methane to synthesis gas and hydrocarbons.

Our purpose is not to be advocates for natural gas, but to point out that neglecting the dynamics of technology can be disastrous in energy planning.

## CONCLUSIONS

In this chapter we examined a number of underlying causes for the blunders in the energy industry. We have tried to be honest, hoping that discussion will prevent a repeat of the mistakes. We tried to show that all sectors contributed to the blunders:

- Academics, scientists, and engineers oversold their ability to analyze and forecast.
- Business leaders searched for simplistic solutions.
- Policymakers chose to follow high-sounding slogans instead of studying the complexities of the issues.
- Activists chose to push their cause without listening to other sides.

We should not underestimate the tactics of talking and not listening—and not thinking. Watch the political debates on television, or listen to a debate on energy in your region, and you will likely find more activists defending positions than seeking solutions. After all, it is easier to argue than to listen and understand!

But understanding is precisely what is needed.

## NOTES

1. A Type I error is the commission of an act that should not have been committed.

2. A Type II error is the omission of an act that should have been committed.

3. It is worthwhile noting that many of the fields in which the United States has achieved leadership in technological quality—such as the electrical generation and distribution system and the telephone system—are essentially cost-plus industries, i.e., regulated monopolies. Even U.S. leadership in the area of commercial aircraft can be traced to the cost-plus military expenditures out of which the industry grew.

4. We pointed out earlier that this may not be true. For example, the security costs for nuclear plants are very high. But it is questionable if they are included in the dispatching decisions.

5. We discuss this technology in more detail in Chapter 6.

6. Rachel Carson (1962) first awakened the general public to a concern that gave birth to the environmental movement in the mid-1960s. Au-

thors such as Hardin (1968) later brought to the public the message of academic writers such as Robert and Nancy Dorfman (Dorfman and Dorfman, 1977). The combination of the popular writers and the academics helped push the U.S. government to create the Council on Environmental Quality and the Environmental Protection Agency through the National Environmental Policy Act (PL 91-190) in 1969. About the same time, the international community formed the United Nations Environmental Program (UNEP), headquartered in Nairobi, Kenya.

7. Of the same size and value as the cathedral of Cologne, West Germany.

8. A gigajoule is a billion joules.

9. Stanford University's Energy Modeling Forum project on U.S. oil and gas supply.

10. This is the first term of Equation 4-1.

11. In refining crude petroleum into useful products, the equivalent of about 10% of the energy contained in the crude petroleum is consumed. In generating electricity from fossil fuels, about twice as much energy is consumed by the generation process itself as is contained in the resulting electricity.

12. The introduction stage (i.e., demonstration plants) reduces uncertainty, not expected cost.

13. *Can*adian *deu*terium.

14. The simplest S-shaped curve is described by the logistic equation. For any parameter $x$ (e.g., performance or market size), $dx/dt = c * x * (k - x)$ where $k$ is the upper limit (saturation value) for $x$, $t$ is the time, and $c$ is a constant. When $x$ is plotted against $t$, it has the form of a symmetrical S curve. The value of $k$ is found at the point where $x$ essentially levels off. We denote $x/k$ as $F$, which is the fraction of saturation value attained by $x$. We plot $F/(1 - F)$ on a semilog plot with $F/(1 - F)$ as the vertical axis and time $t$ as the horizontal axis; the result is a straight line. Delta $t$ is the time required for $F$ to change from 0.1 to 0.9.

15. We have not disaggregated and refined the data into the distances of various flights and the kinds of aircraft needed. A more careful analysis would include this type of refinement.

## REFERENCES

Abernathy, W. J. 1978. *The Productivity Dilemma.* Baltimore: Johns Hopkins University Press.

———, and J. M. Utterback. 1978. "Patterns of Innovation in Industry." *Technology Today* 80 (June–July), pp. 40–47.

Ackoff, Russell L. 1979. "The Future of Operational Research Is Past." *Journal of the Operational Research Society* 30, no. 2, pp. 93–104.

———. 1981. *Creating the Corporate Future.* New York: John Wiley.

Alderfer, E., and H. E. Michl. 1948. *Economics of American Industry.* New York: McGraw-Hill.

Allan, Gerald B. 1975. "Note on the Use of Experience Curves in Competitive Decision Making." 9-175-174. Boston: Harvard Business School.

Awad, A. H., and T. N. Veziroglu. 1984. "Hydrogen versus Synthetic Fossil Fuels." *International Journal of Hydrogen Energy* 9, no. 5, pp. 355–366.

Ayres, R. U. 1987. "The Industry-Technology Life Cycle: An Integrating Meta-Model?" Research report RR-87-3. Laxenburg, Austria: International Institute for Applied Systems Analysis.

Ball, Ben C., Jr. 1979. "Commercialization and the Process of Technological Change." Cambridge, MA: Ball & Associates.

———. 1981. "An Investigation into the Potential Economics of Large-Scale Shale Oil Production." In *Oil Shale, Tar Sands and Related Materials*, edited by H. C. Stauffer. American Chemical Society Symposium Series 163. Washington, DC: ACS Books.

———. 1985. "Management Scientists and Managers: Experiences of an OR-Practitioner with a Critical Interface." *European Journal of Operational Research* 21, no. 1 (July), pp. 17–24.

———, and O. Hammond. 1977. "Recent Proposals for Government Support for the Commercialization of Shale Oil—A Review and Analysis." MIT Energy Laboratory. Report number (May) MIT-EL-77-003. Cambridge, MA.

———, H. Jacoby, L. Linden, M. Adelman, et al. 1976. "Government Support for the Commercialization of New Energy Technologies—An Analysis and Exploration of the Issues." MIT Energy Laboratory. Report number (November) MIT-EL-76-009. Cambridge, MA.

Billington, R., et al. 1983. "Comprehensive Bibliography on Electrical Service Interruption Costs." *IEEE Transactions on Power Apparatus and Systems* PAS 102, no. 6 (June).

Borstein, Daniel J. 1973. *The Americans: The Democratic Experience.* New York: Random House.

Boulding, Kenneth. 1974. "Reflections in Planning: The Value of Uncertainty." *Technology Review* (October-November).

Caramanis, Michael, Richard Tabors, and Fred Schweppe. 1982. *The Electric Generation Expansion Analysis System.* Electric Power Research Institution. Research paper RP-1529. Palo Alto, CA.

Carson, Rachael. 1962. *Silent Spring.* Boston: Houghton Mifflin.

Commission on the Accident at Three Mile Island (Chair, John Kemeny). 1979. *Report to the President.* Washington, DC: Library of Congress.

Deming, W. Edward. 1982. *Quality, Productivity, and Competitive Position.* Cambridge, MA: MIT Center for Advanced Engineering Study.

Dorfman, Robert, and Nancy Dorfman. 1972. *Economics of the Environment.* New York: W. W. Norton.

El Masri, Maher. 1985, 1986. "On the Thermodynamics of Gas Turbine Cycles." Part I, ASME *Transactions*, Vol. 107, October 1985; Part II, *Journal of Engineering for Gas Turbines and Power*, Vol. 108, January 1986; Part III, ASME *Transactions*, Vol. 108, January 1986.

Fisher, J. C., and R. H. Pry. 1971. "A Simple Substitution Model of Technological Change." *Technological Forecasting & Social Change.*

Garvin, David A. 1988. *Managing Quality.* New York: Free Press, pp. 49–61.

Grubler, A., and N. Nakicenovic. 1988. "The Dynamic Behavior of Methane Technologies." In *The Methane Age*, edited by Thomas H. Lee. Dordrecht/Boston/London: Kluwer Academic Publishers.

Hardin, Garrett. 1968. "The Tragedy of the Commons." *Science*, pp. 1243–1248.

Ishikawa, Kaoru. 1985. *What Is Total Quality Control?* Translated by David J. Lu. London: Prentice Hall International.

Jaikumar, R. 1986. "Post Industrial Manufacturing." Working paper. Boston: Harvard Business School.

Lave, L. B., and E. P. Seskin. 1977. *Air Pollution and Human Health*. Baltimore: Johns Hopkins University Press.

Lee, T. H. 1989. "Making Models Matter—Lessons from Experience." *European Journal of Operational Research* 38, pp. 290–300.

———, and N. Nakicenovic. 1988. "Technology Life Cycle and Business Decisions." *International Journal of Technology Management* 3, no. 4, pp. 411–426.

Marchetti, C., and N. Nakicenovic. 1979. "Energy Systems—The Broader Context." *Technological Forecasting & Social Change*, 14, pp. 191–203.

———. 1984. "The Dynamics of Energy Systems and the Logistic Substitution Model." Research Report RP-79-13. Laxenburg, Austria: International Institute for Applied Systems Analysis.

———. 1988. "The Future of Natural Gas." In *The Methane Age*, edited by Thomas H. Lee. Dordrecht/Boston/London: Kluwer Academic Publishers.

Miller, D., and P. H. Friesen. 1984. "A Longitudinal Study of the Corporate Life Cycle." *Management Science* 30, no. 10, pp. 1161–1183.

Nakicenovic, N. 1987. "Transportation and Energy Systems in the United States." Working paper WP-87-1. Laxenburg, Austria: International Institute for Applied Systems Analysis.

OECD Report. 1981. *The Costs and Benefits of Sulphur Oxide Control*. Paris.

O'Keefe, Bernard J. 1985. *Shooting Ourselves in the Foot*. Boston: Houghton Mifflin, pp. 129–173.

Tabors, R. D., and D. P. Flagg. 1986. "Natural Gas-Fired Combined Cycle Generators: Dominant Solutions in Capacity Planning." *IEEE Transactions on Power Systems* PWRS 1, no. 2 (May), pp. 122–127.

Toffler, Alvin. 1980. *The Third Wave*. New York: Bantam Books.

Vernon, Raymond. 1966. "International Investment and International Trade in the Product Cycle." *Quarterly Journal of Economics* (May), pp. 1290–1307.

Weinberg, Alvin M. 1985. "Science and Its Limits." In *Issues in Science and Technology*. Washington, DC: National Academy of Engineering.

White, W. S., Jr., and Ralph D. Dunlevy. 1986. "Managing for Plant Availability." *Public Utilities Fortnightly*, December 11.

Wolfe, Bertram. 1987. *12 Years Later . . . An Update Report on the Nuclear Reactor Study*. San Jose, CA: General Electric Co.

World Health Organization. 1978. "Environmental Health Criteria." In *Sulphur Oxides and Suspended Particulate Matter*. Geneva.

# 6 Lessons Learned

In the previous chapters, we reviewed the blunders and the underlying causes. A large component of human learning is simple trial and error. We try something and fail, and this gives us insight into what to try or not try next. "Good judgment," as former Defense Secretary Robert Lovett is quoted as saying, "is usually the result of experience. And experience is frequently the result of bad judgment." So long as we take seriously the lessons of our mistakes, we are making progress. In recent times, there is encouraging evidence that we have learned from our mistakes. This chapter summarizes the important lessons we have learned in the energy field.

## TECHNOLOGY, NOT RESOURCE DEPLETION, IS THE DRIVING FORCE FOR SUBSTITUTION

The 1973 crisis stirred up a tremendous fear in the United States that we were running out of our most valuable resources. President Carter called the energy crisis "the moral equivalent of war." He turned down the thermostats in the White House, put on sweaters, and passed the Fuel Use Act, forbidding the future use of natural gas in boilers, since it was such a valuable resource. We were running out of oil, gas, and so-called strategic materials among others. The running-out hypothesis—the modern version of the old Malthusian myth—has pervaded the bureaucratic, business, and scientific communities for decades. It has served as a basis for national policy, industrial decisions, investment strategies, and research choices. This myth must be dispelled.

From the dawn of human history, fuel wood, together with animal and farm waste and animal and human muscle power, was the main energy supply. This situation was not altered until the industrial revolution of the nineteenth century. As we discussed earlier, fuel wood was replaced by coal during the last half of the nineteenth century, with fuel wood's share declining from some 70% in 1860 to about 20% around the early 1900s (see Figure 5-20). Fuel wood was abandoned, not because of the threat of resource depletion, but because coal provided an energy source that could do what fuel wood did—and better. Although it was possible (and still is) to operate trains and ships with fuel wood and to use fuel wood for shaft power and electricity, technology advances made it increasingly easier, more efficient, and—most important—cheaper to do so with coal.

By 1910, however, the rapid growth of coal had ceased, with its share of the primary market peaking some ten years later. By the early 1960s, coal had been displaced by petroleum as the dominant fuel, although coal resources were still (and still are) abundant. With the discoveries of petroleum in large quantities—beginning around the turn of the century—industry began to develop a set of oil-related technologies that eventually led to the large-scale refining of oil into a range of products and chemical feedstocks. This abundant supply and low price opened up the market for oil. On the end-use side, refined oil products proved to be far superior to coal for power trains, automobiles and aircraft, generating electricity, locomotives, and for providing residential and commercial heating. All but two of these applications had been first achieved by coal.

Thus, from a historical perspective, energy substitution has been driven by the availability of new technologies that enabled an alternative energy source to better satisfy the demands of society. Oil was cheaper. Therefore, it was better. Oil was suitable for new uses, such as aircraft, making it even better!

Seldom mentioned is that so-called fuel reserves are actually a function of technology and price. The more advanced the technology and the higher the price, the more reserves become known and recoverable. First, as technology progressed, more oil could be discovered and produced economically at a given price. For example, the development of 3-D seismic analysis reduced the cost and increased the accuracy of exploration in certain applications. The development of undersea drilling techniques permitted the exploitation of offshore fields. Second, at higher prices, the utilization of higher-cost technologies becomes attractive. At higher prices, tertiary recovery techniques (such as the injection of chemical agents into the producing zones) become economically feasible for increasing the recovery of oil from existing reservoirs. Consider the history of one field:

> Kern River [oil field] in California was discovered in 1899. After 43 years of production, it had "remaining reserves" of 54 million barrels. In the next 43 years of life, it produced not 54 but 730 million barrels. At the end of that time, in 1986, it had "remaining reserves" of about 900 million barrels.[1] (Adelman, 1987.)

Clearly, forecasts of reserves are subject to uncertainties, both in technology and price. But we can be sure that as technology advances, reserves will increase, simply because of the way they are defined. And, in times of rising prices, reserves will increase, again simply because of the way they are defined.

The reasoning that technology is the engine behind energy substitution led the International Institute for Applied Systems Analysis (IIASA) to forecast—more than a decade ago—that natural gas would be the dominant growth fuel over the next few decades. Advances in natural gas technologies were certainly one of the reasons underlying IIASA's forecast. At that time, these predictions were highly controversial and emotionally charged. They have, however, so far stood the test of time. The consumption of natural gas in the world has been increasing, with the United States being the exception. Until 1987, the Fuel Use Act in the United States drastically reduced natural gas use. With its repeal, it is now legal to use natural gas as a fuel for electricity generation in new boilers. Consumption of natural gas in the United States will most likely increase, in line with trends in the rest of the world.

The implication of this lesson to future energy systems planning is clear: energy systems should be flexible, so that new technologies can be adopted.

## THE CONSEQUENCES OF POOR STRATEGIES CAN BE ENORMOUS AND UNPREDICTABLE

After strategic planning was introduced in GE, its chief executive officer, Reginald Jones, emphasized the importance of simplicity and understandability of strategies by repeatedly telling his managers that if they could not stand on their feet and describe their business strategy in five minutes, then they didn't know their strategy. Such are the characteristics of strategies: simple, easy to understand, and easy to communicate. But two other characteristics of strategies were not emphasized: the consequences of poor strategies are usually enormous, and may be unpredictable. We have already discussed one of the poor strategies and its consequences in Chapters 4 and 5: counting on economies of scale for nuclear power. There are more.

That good strategies need to be simple and easy to understand can be misunderstood by executives in search of simple answers to complex questions. The nuclear example is a case in point. The simple and sometimes successful market-share and learning-curve strategies were not understood in depth by industrial leaders, as we discussed in Chapter 5. There is no learning if every successive reactor is different from the previous one. Yet, during the period when competition in the nuclear power market was severe, businesspeople actually counted on the learning effect (along with nonexistent economies of scale) as a justification to reduce the bidding price. But since no learning was taking place, the assumed learning-curve cost reductions never occurred.

The nuclear industry was not the only one that chose poor strategies. Consider, for instance, the automobile industry. For years, the strategy was to build larger, more comfortable, more luxurious, and more powerful cars for the U.S. market. Only after small cars like Volkswagen started to make inroads into the United States did the auto companies respond by offering a limited menu of small cars. These were merely small, haphazardly designed and built versions of their larger Detroit brethren. The real issue, as we have come to understand, is not large versus small, but efficient and cost effective versus inefficient transportation, and reliable and long-lived versus vehicles with built-in obsolescence. The consequence of this defective strategy is now well known: the loss of significant U.S. market share to foreign manufacturers.

In making strategic plans, we must clearly understand the difference between a plan and planning. A plan is a detailed description of how to implement a chosen strategy. Planning is the process of identifying viable strategic alternatives and the selection of the final choice. For example, if one asked the question whether the United States had a national energy plan in early 1973, the answer, to the surprise of many people, should be yes. Our plan was to depend on low-cost oil from the Mideast, South America, and Africa. The expected prices and volumes were precisely (but not accurately) quantified. We were—and remain[2]—willing to make significant military expenditures to protect that supply. What we did not do was good planning. Good planning must include contingency plans. The contingency that should have been considered in this case was the formation of an effective cartel. In the auto industry the plan was to build large and comfortable cars. The contingency that should have been planned for was a rapid rise in oil prices. But that was not done.

Another example illustrates the importance of contingency planning. We reported in Chapter 5 Marchetti's description of the energy substitution processes, and the resulting forecast: after oil, natural gas may become the dominant energy supply in the world. His forecast was made shortly after the oil embargo when the whole world was concerned with the depletion of two valuable energy resources: oil and natural gas. The Fuel Use Act was passed. The country adopted a policy of depending on coal (and nuclear) for energy independence. A massive R&D program was introduced to extend the use of coal: coal gasification, coal liquefaction, fluidized bed combustion, and a host of other technological options. Coal was king. It was against this background that Marchetti was invited to present his forecasts to GE executives. Halfway through his presentation, one of the senior vice presidents stormed out of the meeting, saying angrily, "I don't believe a word he said."

It is futile for businessmen to engage in a debate on which forecast is correct. The proper questions to ask are:

- If the forecast is correct, what might the consequences be to the company?
- If the potential consequences are severe enough, is a contingency plan (or insurance policy) justified?

The answer to the first question was obvious if Marchetti's forecast was right: GE's steam-turbine business might be in serious trouble, since the bulk of it supplied turbines to large nuclear and coal-fired plants. On the other hand, the stationary gas-turbine business might grow.

The answer to the second question was rather obvious too: GE should have acted to protect its stationary gas-turbine business by investing in R&D and by developing a creative international strategy.

Whether that discussion had a deciding effect on GE strategy, we will never know. In any event, they did pursue gas-turbine market leadership aggressively. In 1986, the large steam-turbine business in the United States, for all practical purposes, collapsed. The gigantic GE plant in Schenectady today depends mostly on spare parts business. At the same time, the gas-turbine business prospered, not in the U.S. market, but in international markets. GE's creative gas-turbine strategy—the manufacturing associates[3] arrangement—propelled GE to the lead position as a supplier of gas turbines.

Strategic planning is a thinking and learning process. It challenges one to analyze the environment, identify the strategic issues and alternatives, design the strategic elements and the programs, recognize the contingencies and opportunities, and prepare plans for them. Most of the tasks require careful and detailed analysis, and are not static. Choosing simplistic answers without the hard analytic work can be disastrous.

## MOST FORECASTS ARE WRONG, THEREFORE, ROBUSTNESS IS A CRITICAL PLANNING REQUIREMENT

The traditional method of planning starts with forecasts of such factors as population growth, growth in GNP per capita, annual housing starts, and the price of oil. Based on such forecasts, strategies are chosen. Unfortunately, many forecasts are wrong, even those formulated by the most famous *and competent* people.

- Robert Millikan, winner of the 1923 Nobel Prize for Physics, said, "There is no likelihood man can ever tap the power of the atom. Nature has introduced a few foolproof devices into the great majority of elements that constitute the bulk of the world, and they have no energy to give up in the process of disintegration." (Cerf and Navasky, 1984, p. 214.)
- Henry Luce, founder and publisher of *Time*, *Life*, and *Fortune*, said in 1956, "By 1980 all power is likely to be virtually costless." (Cerf and Navasky, 1984, p. 210.)
- John von Neumann, Fermi Award–winning American scientist, said about the same time, "A few decades hence, energy may be free—just like the unmetered air." (Cerf and Navasky, 1984, p. 211.)
- George Schultz, former U.S. Secretary of the Treasury, said in 1970, "The world oil market seems likely to be more competitive in the future than in the past because the growing number of production countries and companies make it more difficult to organize and enforce a cartel. This supports the conclusion that the standard price of foreign crude by 1980 may well decline and will in any event not experience a substantial increase." (U.S. Government Task Force, 1970.)

The fundamental problem with forecasts of the future is that they must derive from our understanding of the present, which in most cases is incomplete. Large, dynamic systems—the economy, financial markets, energy demand—are driven by endogenous and exogenous variables whose causal relationships we only dimly understand. That is why your barber's opinion of the direction of the stock market is usually no better or worse than that of Wall Street analysts.

We pointed out earlier that it is futile for decision makers to spin their wheels in debating whose forecast is correct. Instead, the proper question to ask is, What does it mean if a seemingly unlikely forecast turns out to be right? Contingency planning is a way of accounting for this eventuality as well as a method for assisting in formulating a basic, robust strategy.

An example of this point is the choice between large and small plants in the electric power industry. We have already questioned the validity of the concept of economies of scale when considering nuclear plants. For our discussion, let us accept the concept for other generating units such as coal, oil, or gas. In the first half of 1974, a few months after the Arab embargo, the electric utility companies in the United States ordered 140 gigawatts of new generating plants. During the second half of the same year, the utility companies started to postpone the orders, because the traditional load growth of 7% per year did not appear to be true anymore. Over a period of six to eight years, most of the orders were postponed or cancelled, at great expense to the companies. What do these actions mean to the utility companies and what could be their strategic choices?

A generating plant, under ideal conditions, would have a particular expenditure pattern (see Figure 6-1). Once the decision is made

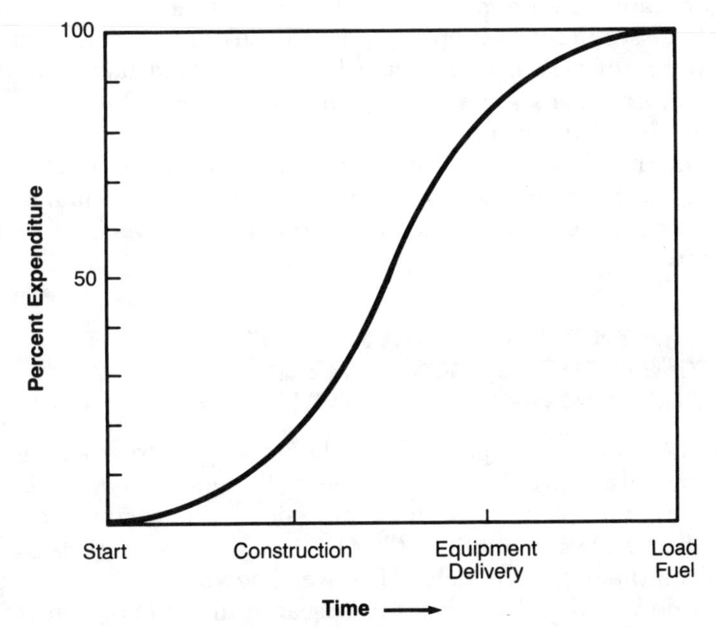


**FIGURE 6-1 Typical Construction Expenditure Pattern for a Nuclear Plant.**

to build, the expenditure begins. In the beginning, it is relatively low; most of the expenses are for engineering design and planning. Shortly after that, the rate of expenditure accelerates as site preparation and ordering of equipment require significant resources. Suppose now the load growth forecast fell to less than the one made before the decision to order the plant. What are the choices?

The CEO might believe that low load growth is a temporary disturbance and growth will soon resume and choose to ignore the forecast and proceed as planned. If the judgment turns out to be wrong, the costs of postponement and cancellation at a later date will be much higher. Alternatively, the CEO might believe that in the near term the low growth will persist, but in the long term the growth will recover. The choice would be to postpone the plant construction and accept the penalty of a higher plant cost. Or, the CEO might decide that the low growth will continue. The choice would be to cancel the plant and absorb the losses. Examples of all three choices were seen in the years after 1974.

Suppose the CEO had said before the decision was made, "I don't know whether to believe the forecast or not. If it's too optimistic, what does it mean to me if I ordered the plants the forecast calls for? Or, what if I hedge my bets and order smaller plants, spacing them apart in time, and take our chances that we may lose some money because of economy of scale, but protect ourselves against the horrible losses in interest on capital and sunk costs?" Questions like these were not asked. If they had been, we would have found the slightly more expensive and less optimum decision to be more robust than the so-called optimum solution.

For the planning of future energy systems, experience tells us to make contingency planning a central part of decision making and to search for robust strategies, not the one that today's favorite forecast shows to be optimum.

## THE TASK OF ANALYSTS IS TO LAY OUT THE OPTIONS, NOT TO TELL DECISION MAKERS WHAT TO DO

In Chapter 5, we pointed out the importance of knowing how to use models. Let us talk about one of the good ways. Prof. Song Jiang, the chairman of the State Commission for Science and Technology in the People's Republic of China, is a highly accomplished scientist and mathematical modeler. He is well known in the field of control theory and was the director of the Research Institute of Control and Informatics before becoming commission chairman. In the latter capacity he is principally responsible for the reform of the scientific system in China. At the beginning of a speech he delivered in 1988 he said:

> Some years ago, when I was still wholly devoted to research, I too felt that systems analysts could one day unequivocally formulate all problems of the world with mathematical equations. It seemed to me that political leaders were often caught in dilemmas and troubles only because of their ignorance of systems analysis. I was even convinced that as long as we could discover the key control variable of any problem under consideration, we would be able to solve it readily at once. Of course, if that were true, we would be living in a much better world today.
>
> However, since I began to handle government affairs, I have gradually come to realize that my idea about the omnipotence of systems science was at least partially wrong, if not all. We must admit frankly that many important issues are simply beyond quantitative description. The mathematical tools provided by systems science today fall far short of being able to describe all social phenomena, let alone such complicated factors as political movements or trouble-making students, either in China or elsewhere. I came to the understanding that many serious problems in society as well as in science call for [a] combination of scenario analysis and systems analysis. In many cases, one can do better by using simple logic and popular verbal expressions in persuading people to do something. This, however, was something I used to underestimate in the past. (Song, 1988.)

It is clear from Song's speech—delivered at IIASA, where such famous systems scientists as Nobel laureates Kontorovich and Koopmans, as well as Danzig and others worked—that lessons have been learned. Song's points can be illustrated with a few examples.

## An R&D Planning Methodology for the Electric Power Research Institute (EPRI)

Supported by the majority of electric utility companies in the United States, EPRI contracts and manages research and development for the entire industry. It has an annual research budget of about $300 million and supports, in total, more than a thousand projects. With many projects on its plate, EPRI managers face the challenge of organizing and prioritizing their efforts.

In the early 1980s, opinions within EPRI were divided on how to approach the challenge. One school of thought leaned toward developing a comprehensive, analytical strategic-planning model. The other proposed to decompose the entire issue into a number of relatively independent or, in mathematicians' language, orthogonal questions, so that each one could be studied on its own terms. Although both approaches were pursued, the first was never completed. The second approach enjoyed more success. In 1988, the National Science Foundation of Norway (NTNF) formally adopted the methodology for its planning work (for more detail, see Lee, Fisher, and Yau, 1986).

Searching for orthogonal functions is nothing new for scientists, engineers, and others who solve complex mathematical problems. Such functions preserve their identity and significance when combined with other orthogonal functions.[4] EPRI's management problem led to the identification of five orthogonal questions, each related to a specific issue. Management then developed checklists of relevant factors for each issue and means for displaying each issue's key aspects for managerial consideration.

Three of the five issues, and the corresponding questions that define them, evaluate the merits of individual R&D projects:

- Project strength: Are the individual projects worth doing?
- Project timing: Is progress toward commercialization in step with technical and economic readiness?
- Project fit: Does the project fit the corporate objective and is the funding pattern correct?

The other two issues evaluate the merits of an organization's overall R&D program:

- Program responsiveness: Is the overall R&D program responsive to corporate strategy and objectives?
- Program robustness: To what extent will program strength be maintained in the face of discontinuities in the social, economic, or political environment?

No formal proposal was made for combining the answers to the five questions into a single criterion for R&D planning and budgeting. Management's unique characteristic, including its broad intuitive judgment, is required and utilized for subjective integration of the trade-offs between the functions. Thus, the tasks of integration and decision making were left to the operating management, with the assistance of its planning staff.

EPRI managers were provided with a comprehensive framework to facilitate their decision making. For each issue, they would evaluate projects or programs from two different perspectives and display the resulting evaluations in matrix form. In evaluating project strength, for example, project achievability (related to the quality of the technical work, the probability of technical success, and the probability of market success) was one dimension of the matrix. Potential value (related to a number of economic and market factors) was the other. The purpose was to provide a comprehensive checklist of factors to be considered and to display the results of evaluations consistently. The suggested checklists were considered illustrative rather than definitive. An organization that adopted the methodology for its R&D management process would be expected to modify and tailor the checklists to fit its managerial practices and requirements.

The use of a two-dimensional display for strategic-planning purposes deserves comment. It began when GE formally introduced strategic planning as a management function. The first major task was to decide how to allocate financial resources to the approximately sixty strategic business units (SBUs). For that purpose, two measurements were identified for each business: industry attractiveness and business strength. In a two-dimensional plot, each business is represented by a point. The resource allocation decision for that business depends on the location of the point. When the approach was first introduced, operating managers reacted quite negatively, for it fundamentally challenged their knowledge of the business. In reality, planners can never challenge the detailed knowledge and judgment of operating managers. Thus, arguments on where individual businesses belong on the two-dimensional plot were difficult for the planners to win. Nevertheless, the two-dimensional display does allow the planners to argue the relative positions of different businesses without having to rely on fine details.

Faced with the issue of relative positions, the operating people had to be objective. In the end, sensible overall resource allocation strategies were derived. Since that time, many organizations have applied the approach to other management issues such as life cycle management.

It is interesting to note that, in the original GE approach, many factors were considered for each measurement, but there was no defined way to combine the analysis for different factors into a single bottom line. That actually made it easier for the final consensus, although the process involved a great deal of dialogue and interaction.

In the EPRI study (Lee, Fisher, and Yau, 1986), we had the same experience. It will be illustrated with two examples: the first and the last orthogonal issues.

**Project strength.** Project strength is the basic issue when assigning priority to projects. The questions here involve the extent to which the ultimate goal of the project is actually achievable, and the potential value of success. Two aspects of achievability must be considered: technical achievability (will it work?) and market achievability (will it compete?). (See Tables 6-1 through 6-3.)

When faced with the necessity of ranking projects, people tend to become engrossed by scoring systems, arguing interminably about which to use and which not to use. In our view, these are side issues. *Any* relevant, well-considered scoring system, whether numerical, ordinal, or nominal, is equally useful. It is only necessary that the scoring system be applied to the evaluation of every project with an even hand and good judgment.

Estimating potential value, the second dimension of project strength, is unavoidable even though difficult because it depends on a variety of factors. As an aid to estimating, we suggested that five aspects of value be weighed individually: direct economic value, social (including environmental and other indirect values), uniqueness, urgency, and timing. Results of the five assessments would then be combined into a single overall estimate by whatever means best suits the style of the institution involved.

Having evaluated project achievability and potential value for a number of projects, the results of the evaluation can be displayed in a matrix for review by general management. Sixteen projects funded by the Energy Management and Utilization Division of EPRI

**TABLE 6-1 Sample Checklist for Technical Achievability**

Technical achievability is determined by the difficulty of the work to be done and by the availability of the required resources. Consider the following questions:

How difficult and demanding is the job?

- What are the technical obstacles?
- How surmountable do they appear to be?
- What unexpected problems have arisen?
- Are milestones being met?
- How likely are schedule slippage and cost overruns?

Are the required resources available?

- What skills are required? Available?
- What facilities are required? Available?
- Is contractor or subcontractor performance acceptable?
- Are financial resources available?

*Source:* Technology Assessment Group. 1982. *Development of Methodologies for Planning of Research and Development Programs, Phase II—Final Report.* Schenectady, New York. Submitted to the Electric Power Research Institute.

**TABLE 6-2 Sample Checklist for Market Achievability**

Market achievability is determined by the competitive economic merits of the project and by the ability to transfer technology from the R&D team to the manufacturing and marketing organizations. Consider the following questions:

What is the commercialization plan?
- Potential users? Market size?
- Projected cost? Competitive?
- Competitive developments?
- Patent protection?
- Targeted manufacturing organization?
- Targeted marketing organization?
- Barriers to commercialization?

Technology transfer plan?
- Publications?
- Pilot plant or demonstration?
- Transfer of personnel?
- Experienced manufacturing organization?
- Experienced marketing organization?
- Fit to current practices?
- Barriers to technology transfer?

*Source:* Technology Assessment Group. 1982. *Development of Methodologies for Planning of Research and Development Programs, Phase II—Final Report.* Schenectady, New York. Submitted to the Electric Power Research Institute.

**TABLE 6-3 Sample Scoring System for Overall Project Achievability**

On the basis of answers to questions on technical and market achievability, overall achievability can be scored from 10 to 0.

| | |
|---|---|
| Eminently achievable | 10 |
| Average achievability | 5 |
| Not basically achievable | 0 |

*Source:* Technology Assessment Group. 1982. *Development of Methodologies for Planning of Research and Development Programs, Phase II—Final Report.* Schenectady, New York. Submitted to the Electric Power Research Institute.

were evaluated by their project managers. Even though the projects were evaluated by their own managers, each striving to make his or her own project look good, the discipline of evaluation according to standardized measurement criteria under the guidance of the division planning staff led to a wide range of scores. (See Figure 6-2.) Three other orthogonal issues were studied in a similar way.

**Project robustness.** The last orthogonal issue, robustness, required even more interaction with decision makers. The concept of

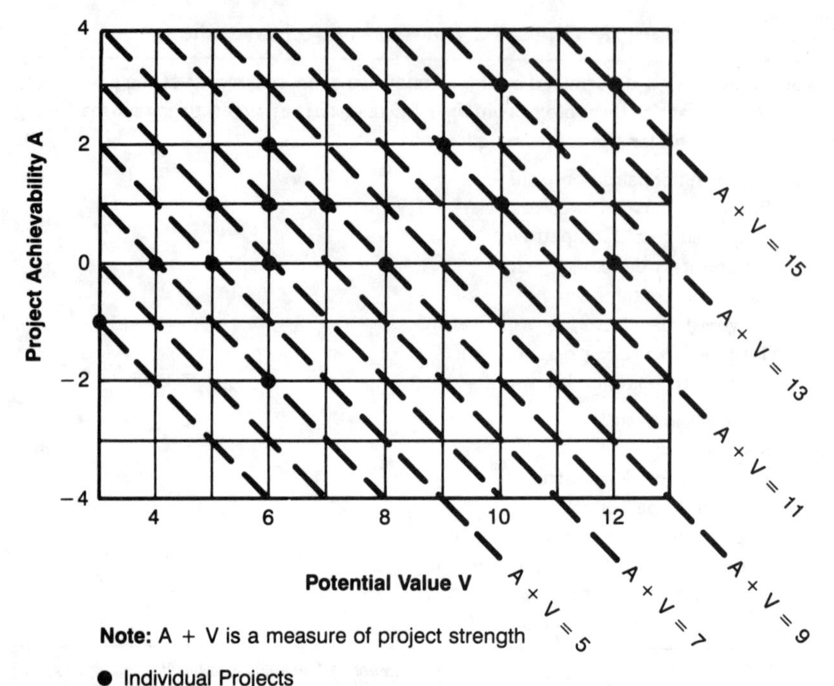


*Source:* T. H. Lee, J. C. Fisher, and T. S. Yau. 1986. "Is Your R&D on Track?," *Harvard Business Review*, January-February, p. 34.

**FIGURE 6-2 Project Strength Matrix for 16 Projects.**

robustness provides a way for managers to anticipate and safeguard their programs against the economic, social, or political shocks that can radically affect the climate for commercialization. But the managers must decide what are the shocks they want to protect themselves against. The planners convened a meeting of one of their utility advisory committees, which was asked to draw up a list of potential environmental shocks. Each member of the committee was asked to estimate the probability of occurrence for each shock. (See Table 6-4.) The sum of all probabilities is 3; that means the committee expected about three of the thirteen shocks would actually occur in the next dozen years.

The next step was to assess the impact of each one on each project. A shock can influence project strength, timing, or fit. The degree of overall impact was again left to the judgment of the committee members. The results of the evaluation can be displayed in two different ways:

**TABLE 6-4 Potential Environmental Shocks to Electric Utility Planning**

| | *Environmental Shock* | *Estimated Probability of Occurring by 1995* |
|---|---|---|
| 1 | Significant tightening of outdoor air-quality standards | 0.38 |
| 2 | Severe limitation on raising of capital funds | 0.29 |
| 3 | Load growth twice (or more) currently projected rate | 0.28 |
| 4 | Serious deterioration of regulatory climate | 0.28 |
| 5 | Political curtailment of oil imports (one year or more) | 0.26 |
| 6 | Severe restrictions on use of water in power generation | 0.25 |
| 7 | Significant tightening of indoor air-quality standards | 0.22 |
| 8 | Unlimited natural gas at constant real prices | 0.21 |
| 9 | Oil price rises twice as fast as projected | 0.21 |
| 10 | Decrease in real price of oil (ten-year duration) | 0.19 |
| 11 | Load growth half (or less) currently projected rate | 0.19 |
| 12 | Severe constraints on operation of nuclear plants | 0.14 |
| 13 | Significant lessening of institutional constraints against nuclear plant construction | 0.11 |
| | | 3.01 |

*Note:* Estimates are as of 1983.

*Source:* T. H. Lee, J. C. Fisher, and T. S. Yau. 1986. "Is Your R&D on Track?," *Harvard Business Review*, January–February, p. 42.

- a plot that shows the overall assessment of all projects against all probable shocks, or
- a similar plot for all projects against individual shocks.

EPRI has a rather elaborate committee system to plan and review its research activities. The proposed evaluation system was tried in an interactive way with several committees in one division. The results were quite satisfactory.

The example leaves us with several lessons:

- Decomposition of a complex problem into orthogonal issues makes it easier for decision makers to understand and work with analytic methodology. It is important to point out that this does not violate our arguments for the systems approach. In this case, the interaction among the orthogonal issues was left to the decision makers in an interactive way.
- Detailed analytical approaches to scoring and combining different aspects of a problem should be left to the decision makers.

## Reducing Variables for Decision Makers

In the late 1970s, the Consolidated Edison Company of New York City asked the MIT Energy Laboratory to perform a strategic-planning study. The central issue was facility planning from the economic, rate-setting, and ecological perspectives. How should Consolidated Edison choose from among its many options, which included adding pump hydro plants, converting existing oil-fired plants to coal, and adding scrubbers to coal plants. At the very beginning management acknowledged four central points:

- A large number of alternate possibilities;
- Massive uncertainty;
- Multiple attributes; and
- Multiple decision makers, often with different goals and outlooks.

The MIT Energy Laboratory developed a simple, straightforward and practical approach involving four steps.

*Step 1*. Define the problem in terms of the options available to the utility, and how they can be combined into plans. Identify the uncertainties. Identify the attributes that should be used to evaluate the plans in the presence of alternate futures.

*Step 2*. Evaluate the attributes of a representative sampling of the plans and futures. This step is not too difficult, especially if the necessary computer simulations are available, but it is very tedious. A large number of computer runs are required. But there are methodologies for interpolating between runs, so that a large database can be obtained from a reasonable number of actual computer runs.

*Step 3*. Throw out all inferior plans, namely, those that are dominated by another plan. (See Figure 6-3.) For example, plan $P_1$ dominates plan $P_2$ if all the attributes of $P_1$ are better. Only those plans that are on the trade-off curve are of interest. They are called the decision set.

Because the future is uncertain, dominance is defined in a probabilistic sense. A plan is judged to be good if it puts us on the frontier in each of the possible futures (contingencies).

*Step 4*. Do a diagnostic analysis on all remaining plans. Here the plans of the decision set are examined. They are the plans on the trade-off curve. The examination leads to diagnostic analysis of the following types:

- Critical and Noncritical Uncertainties. Are some uncertainties especially important or unimportant to a particular plan, set

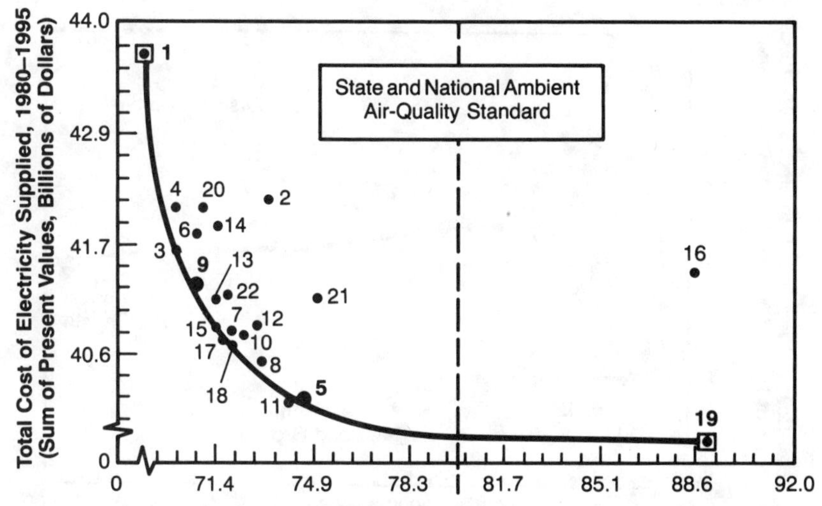


*Source:* D. C. White et al. 1981. "Strategic Planning for Electric Energy in the 1980s for New York City and Westchester County," MIT Energy Laboratory, Cambridge, MA.

**FIGURE 6-3 Total Cost (1980–1995) vs. Peak 1995 Sulfur Dioxide Concentrations for the Exploratory Scenarios.**

of plans, or overall decision set? For example, the plans on the trade-off curve can vary enormously under one uncertainty, such as load growth. (See Figure 6-4.) Among the Con Edison plans, all of the attributes are very sensitive to the uncertainty, but the choice of the plans themselves does not change. The same plans lie on the trade-off curve for low-load growth and for high-load growth. Hence, in this case, load growth is a noncritical uncertainty, a surprise but a strategically important finding.

- Critical and Noncritical Attributes. Some attributes are more important than others. The Consolidated Edison study originally involved a list of thirteen attributes. Three of these were quantity of imported oil, present value of total costs, and the rate paid by the utility customers in the final year being studied. All three were found to be linearly related. Relative to determining which plans were worth considering, only one was required.

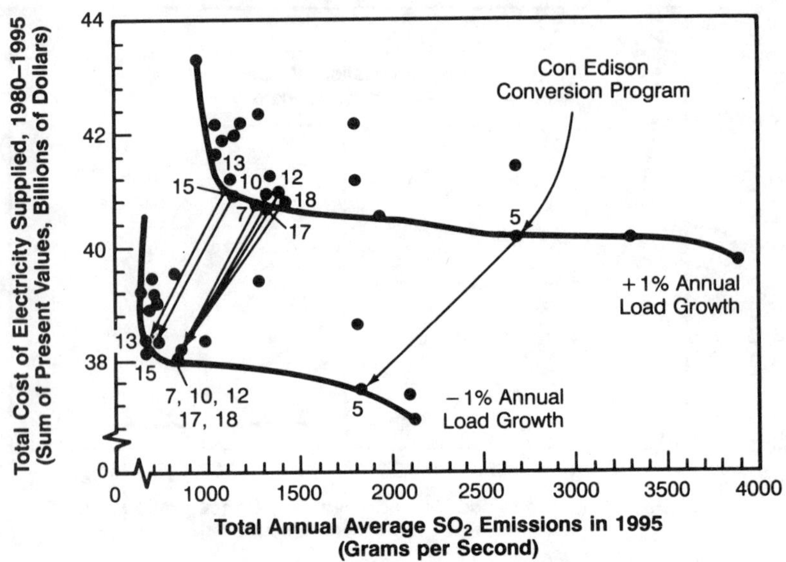


*Source:* D. C. White et al. 1981. "Strategic Planning for Electric Energy in the 1980s for New York City and Westchester County," MIT Energy Laboratory, Cambridge, MA.

**FIGURE 6-4 Effect of Change in Load Growth on Total Cost (1980–1995) and 1995 $SO_2$ Emissions for the Exploratory Scenarios.**

- Dominance, Robust and Inferior Plans, and Options. Are there any plans that almost always look good, independent of which future occurs? Are there options that are dominant or inferior? In the study used as an example here, one particular option, purchases of electricity from an outside source, always turned out to be a part of any plan on the trade-off curve for any future. That is, it was a dominant option. On the other hand, a particular type of plant construction was always inferior. Similarly, there were certain classes of plans that always lay on or close to the trade-off curves—independent of what futures were being considered. These were considered robust plans.

One additional aspect of the study deserves mention. Consolidated Edison recognized that the decision concerning the electric power company serving the City of New York could not be made by the company executives alone. An advisory committee was formed, consisting of city officials, environmentalists, labor unions, industrialists, and so forth. The group met several times a year to hear detailed presentations and discuss alternatives. As a result, when the study was completed, all concerned had been a part of the process and understood the conclusions.

The process is not designed to yield a final optimal plan. Rather it is a process of discarding the bad or inferior alternatives. It is a process of elimination to find those that are good across a wide range of uncertainties—in a word, robust options. The process brings the work of the analyst to the decision maker. The decision makers can then do what they are supposed to do, make decisions by considering the attributes of alternative options.

The lessons we can learn from this example are:

- Strategic planning is less concerned with finding optimum solutions than with exploring the ranges of uncertainties and trade-offs.
- It is important to involve opinion leaders and decision makers early and all along the way.

## Interactive Decision-support Systems

One of the frustrations of IIASA management and scientists was that in spite of the outstanding analytical work done in the institute, its application in the policy community had not been completely satisfactory. A continuing problem for IIASA directors has been how to make IIASA work more useful. On the one hand the institute's analytic work is unquestionably excellent. On the other, one could not help noticing the limited application of IIASA models. In searching for a remedy for the situation, IIASA found an important principle: Interactive decision-support systems are far more attractive to decision makers than are large computer models. Decision makers are hungry for good methodology and tools, not canned solutions. This goes to the heart of the problem described by Song Jiang of China. Of course, IIASA is not alone in this discovery. Many organizations are experiencing the same thing and interactive decision-support systems are now finding useful applications to major problems around the world. To name a few:

- Water management for lignite operation in the German Democratic Republic and the agricultural industry in the Netherlands.
- Acid rain control in conjunction with the United Nations Economic Commission of Europe (ECE).
- Management of hazardous waste materials for the Commission of European Communities.

- An expert system for the development of coal resources in the Shanxi province of the People's Republic of China.
- Management of chlorine in the Netherlands.

In these cases, the system integrates a number of databases containing relevant information with several existing simulation models. The individual modules are of proven validity, so that decision makers can feel confident in their application. The modules are interconnected so that the inputs are consistent and the output of one module can be part of the input of the next. In addition, the user interface is designed for maximum ease in a work environment through high-resolution color graphics and user friendly menus.

The IIASA acid rain project produced a widely accepted interactive system (Alcamo et al., 1984). The model system was codesigned by analysts and potential users. It was of modular construction and consisted of a series of linked submodules. Submodels were as simple as possible and based, when feasible, on more detailed models or data. They were made more complex only if necessary and only in conjunction with potential model users. The model had interactive input and clear graphic output and presented a temporal picture of the problem.

Analysts have learned what they should do: offer help by providing methodologies and tools for decision makers, not prescribed answers.

## MEASURE THE RIGHT THING

We discussed in Chapter 5 the importance of measurements and the disastrous consequences that can result when wrong measurements are followed religiously. We also indicated that we may be on the verge of repeating some of those mistakes. Therefore, this is a lesson that deserves reemphasis. When we say measurements may be wrong, we do not mean that we did a poor job in measuring, that something is really eleven inches long, and we find it to be twelve inches. Rather, it is that we are measuring the wrong thing; we have measured the length when we should have applied our yardstick to the width. More likely still, we should have been concerned with measurements in all dimensions, but we only paid attention to length.

The emphasis on the levelized or busbar cost of electricity (COE) is a case in point. Throughout much of the energy crisis we focused our attention on developing new technology that could compete on a kWh-by-kWh basis. We neglected to consider that our electric power systems were just that, systems, with significant capital stock in place. Any new technology needed to be integrated with the existing system and provide energy into that system. We also tended to ignore the other attributes of the technology. What of the environmental concerns, the emissions of sulfur dioxides or carbon dioxide? What of the problems of delivery of the energy, the difference between trainloads of coal and a pipeline of gas? What of the concerns for siting of facilities, the convenience of smaller, more modular units with smaller footprints on the countryside? What of the needs for what may be, in the long term, the scarcest of our resources, water? These dimensions were ignored in the COE. COE is not the wrong yardstick; it is simply not the only one.

Perhaps we should examine the measurements we use in real life in much the same way as we examine the rules in sports. We should think about every measurement very carefully, because we know once they are put in place, we will not or cannot afford to debate or change them during the game. We also know that the rules make the difference. If dunking were not allowed in basketball, Michael Jordan could not be the great player he is. But who should examine the rules? In sports, we have commissions for such a task, even on an international scale. Who bears such responsibilities in the energy and environmental arena?

For example, a significant part of the problem of the U.S. inability to compete in world markets can be attributed to our measurement system, which is based largely on short-term performance. Corporate managers are held accountable primarily for quarterly results, current dividends, and near-term prospects. The game thus defined, it is neither possible nor prudent to work toward long-term goals, no matter how important or even crucial they might be.

On a much larger scale, the entire question of a political system is mostly a question of measurements. John Locke had a different system of measurement from that of Lenin. Deng Xiao Ping's measurements (primarily economic) are certainly different from Mao Tse-tung's. In a democratic society, it is everyone's responsibility to question the validity of every single measurement we have to live with. The ecology movement is an excellent example of a popular movement that has effectively changed the rules of the game, first in the West and increasingly in the socialistic countries.

Two specific examples illustrate the general problem of measurement. The CEO of a large multinational corporation was questioned at a recent university lecture about his decision to divest what most considered to have been that company's core business because the market had been shrinking. Some in the audience argued that he was unpatriotic to let foreign suppliers take over a business critical to the nation's economic system. Others raised the question of whether or not he should continue to run that business at a loss only to keep the production capability in the United States. The issue is, What are the rules of the game he is playing? The idealist's rules that suggest keeping the technical capability in the United States are not those used by his board of directors in assessing his performance. The second example is equally difficult. What role should or does R&D play in a company subject to merger or buyout? As one CEO recently commented, what CEO would invest in long-term R&D if the problem he or she faces is the end of the company through an imminent buyout? The conclusion one comes to in discussions like this has to be that it is not the CEOs who have bad intentions. Rather, their hands are tied by the rules we have set for the financial and industrial games in this country.

## DO NOT CONFUSE THE SYSTEMS APPROACH WITH SYSTEMS ANALYSIS

We are becoming increasingly aware of the complex interrelationships between the activities of man and those of the remainder of the physical and environmental system. Throughout the process of learning we find new examples of our inability to look at the system as a whole. We can no longer limit ourselves to examining phenomena in isolation, such as sulfur emissions from power plants and nitrogen oxides from automobiles: instead, we must consider the ecological system as a whole. According to the 1985 *Annual Report* of IIASA,

> Until recently, the occasional collision between efforts for societal well being and the environment's capability to sustain these was a discrete, local, and relatively straightforward affair. But we have entered a new era of increasingly complex patterns of environmental and societal interdependencies that could compromise the sustainable development of the biosphere. Government and industry planners now confront tradeoffs between development objectives and environmental concerns in the face of significant scientific uncertainty and minimal social consensus. Acid deposition, soil erosion, stratospheric ozone depletion, greenhouse effects, and the like represent tightly coupled geochemical syndromes in need of integrated analysis. They are linked through development policies that are the common causes of many environmental perturbations. The critical nature of such issues often reflects unanticipated long-term

> consequences of development activities undertaken for their short-term benefits.
>
> The time is ripe to expand efforts for understanding the interaction of the development activities and the environment. A unified view of the structure and functioning of the biosphere is emerging based on increased knowledge of major biogeochemical cycles and their relations with one another and with the global climate system. (IIASA, 1985.)

In the IIASA project "The Sustainable Development of the Biosphere," two committees were organized for each specific issue: a scientific committee and a policy committee, with distinguished members drawn from both communities. The committees interact with one another so that scientific issues can be examined simultaneously with policy alternatives. This is a significant departure from the past approach of constructing the models first, then interacting with the policy communities. In both cases, the recognition of the systems aspect was clear, but the later one emphasized the synthesis and design.

For future energy systems planning, we should leave enough room for design, not *just* analysis.

## UNDERSTAND TECHNOLOGIES AND MANAGE THEM ACCORDINGLY

In Chapter 5 we discussed the need to assess technologies realistically, to be sensitive to the dynamics of commercialization, and the life cycle phenomena, and to avoid excess capacity problems in the mature phase of the life cycle. In the case of technology assessment, the most important thing is to choose the proper assessment criteria. We have discussed the problem of using the cost of electricity as the only criterion for the comparison. Intangibles and contingencies should have a significant role. A recent case illustrates the point.

The $300 million Coolwater Project in Southern California sponsored by the Electric Power Research Institute, the Southern California Edison Company, Texaco, GE, and others was intended to demonstrate the technical and economic feasibility of an integrated gasifier combined cycle system (IGCC), which gasifies coal to create an environmentally acceptable gas to be burned in a gas-turbine combined cycle. The initial economic selling point of the project was the levelized cost of electricity, which was supposed to be 10% to 15% lower than conventional methods of generation. The plan was to negotiate with the California authorities a contract for the sale of electricity from that plant at a price allowing the partners to recover their investment in a relatively short time. For GE this was a reasonable investment, since GE was by far the leading supplier of combined cycle systems. But the internal management was not persuaded by the economic forecast. It looked beyond that. First, it organized several design review teams, one to assess the gasifier technology, one to assess the material behavior, and one to assess the reliability of the plant. The conclusion after all the careful studies was that there were enough technical uncertainties to question the validity of the COE calculation. Nevertheless, the likelihood of technical success was good. Second, it made calculations on the sensitivity of costs of different technologies to environmental regulations. It was found that IGCC was insensitive to the introduction of more stringent environmental regulation, whereas the conventional technologies experience rapidly rising cost under these conditions. That consideration alone carried more weight in GE's decision to join the project than did the COE calculations.

Events in the energy world such as the discovery of "fusion in a jar" have brought these points home even more strongly. Heralded by many as the savior of the future, the question is, What is it? The simple answer is that as of this writing no one knows what it is, though we can say more clearly what it is not. It is not technology, it is science. Maybe, and we hope some day, it will grow up to be technology and have a role in the energy and environmental future of our globe. Until it matures to a technology and develops to a commercially viable and integratable technology it is only interesting science that holds the same dream that many of the solar technologies hold, that fuel cells hold, and that MHD holds: a glimmer and a hope that someday it will be cost effective.

The lesson we have learned should be clear: understand the technology and its life cycle, and manage it through the process. Do not try to force it through.

## QUALITY PAYS AND CLEANER IS CHEAPER

The issue of international competitiveness has brought the quality question to center stage. Management in the United States is coming to realize that higher-quality products do not necessarily mean higher total costs. We are learning to build in quality to reduce the number of failures and returns under warranty, and to increase customer acceptance.

The issue of "cleanliness" is, in many ways, closely associated with that of quality. By *cleanliness* we mean the amount of environmental residual produced in the manufacture of products. Today's views on cleanliness are similar to the views on quality thirty years ago. Cleanliness is a long-term problem and one whose benefits to today's businesses are intangible. Only the few companies that find themselves on the wrong end of environmental lawsuits are feeling the pain of past actions: Hooker Chemical at Love Canal and the major chemical and pharmaceutical firms being challenged for land dumping of hazardous wastes. These are the blatant—and traceable—examples in which the problem has been returned to the doorstep of the polluter.

The problem will be returning to more and more corporate doorsteps as time passes, for there are more environmental residuals, water pollutants, and air pollutants like sulfur dioxide and even carbon dioxide. Can manufacturing that requires significant inputs of energy be carried out so that dangerous emissions are not released into the environment? The answer is yes. But the question of costs has not yet been fully answered, as we will show in Chapter 7.

Traditionally, studies on ecological systems have been based mostly on an evolutionary paradigm. The models are usually surprise-free, as pointed out by Brooks (1986). They depend on the incremental unfolding of the world system with parameters derived from a combination of time series and cross-sectional analysis of the existing system. Of course, this is precisely the reason why most forecasts are wrong. In real life, surprises are the rule rather than the exception. Brooks divided surprises into three types:

- Unexpected discrete events, such as an oil shock, nuclear power plant accident, and war.
- Discontinuities in long-term trends, such as the stagflation phenomenon in OECD countries in the 1970s. To this category we should add the gradual loss of market share of oil to natural gas as predicted by IIASA.
- The sudden emergence into political consciousness of new information, such as the relation between fluorocarbon production and stratospheric ozone, the deterioration of central European forests apparently from air pollution.

We have already pointed out that blunders in the energy field are mostly because of our inability to handle surprises of the first kind. The importance of contingency planning for surprises of the second kind is clearly described in the gas turbine case. In ecological areas, our challenge is how to manage surprises of the third kind. This is one of the primary objectives of the IIASA project on the ecologically sustainable development of the biosphere. It will determine the extent to which methods, models, and concepts can be advanced beyond surprise-free analysis to a more realistic treatment of the interaction between development and environment. Many instances of surprises in ecological systems and how they interfered with surprise-free management are described by Clark (1986) and will not be repeated here.

Although we can assert that "cleaner is cheaper," we have yet to prove it. Some manufacturers are moving in this direction. We as a society are moving there because we are more conscious of environmental residuals in our personal lives and recycle more of our waste. Only increased experience and experimentation will demonstrate the truth of our assertion in absolute terms, though its truth in relative terms cannot be denied.

## DO NOT OVEREMPHASIZE SCIENCE AND DE-EMPHASIZE ENGINEERING

Of all the lessons we include in this chapter, this one will probably lose us the most friends: Do not overemphasize science and underemphasize engineering. But the truth is, the policymakers in this nation have only occasionally understood the difference between science and engineering. The president has a science adviser, but *not a technology or engineering adviser.* The possibility that the science adviser may be ignorant about the engineering aspects of converting science into something practical never seemed to worry anyone, until the recent problem of international competitiveness surfaced. But if we reconsider the blunders in the energy field, we must acknowledge that engineering problems—including engineering economics (not economic theory)—were responsible to a significant degree. Let's list a few.

- We thought that the most important issues in nuclear power plants were nuclear physics problems. We never considered chemical engineering problems until it was too late. Looking at the outages, forced or planned, one can easily see how few of these were related to science and how many were directly related to systems engineering. Knowing that it worked in the lab, or even in a submarine, was not enough.
- We never considered the engineering and engineering economics aspects of fusion until Larry Lidsky, a leader in MIT's fusion research, published an article entitled "The Trouble with Fusion" (1983), in which he said, "The scientific goal of the fusion program turns out to be an engineering night-

mare. Even if the fusion program produces a reactor, no one will want it" (1983, p. 34). Lidsky may have created many enemies in the fusion community, but he was speaking the truth. The engineering and engineering economic aspects were not understood, principally in the policymaking community.

- We are probably witnessing another blunder. The discovery of high-temperature superconductors is undoubtedly one of the most significant scientific discoveries since the transistor and DNA. But the forecasts for practical commercialization, we now see, have all the hallmarks of past mistakes. Forecasts have not paid adequate attention to either the engineering or the dynamics of the commercialization of technologies.
- The possible discovery of what is called cold fusion or test tube fusion is all the rage at this writing. Were it successful it would undoubtedly be hailed as the savior of the environment and our energy future. In the long run it might help. But in all the print devoted to this subject, almost nothing has been said about the engineering issues needed to turn the science of fusion into something that comes out of your wall socket. If recent cold fusion discoveries pan out, which seems unlikely, they will have to travel the long and bumpy course of commercialization of new technologies, as we discussed in Chapter 5.

Scientists must help the public understand that they face limitations in design, commercialization, practical economics, operation, and maintenance.

## GOVERNMENT-SPONSORED R&D PROJECTS IN AREAS WHERE THE GOVERNMENT IS NOT THE USER OF THE RESULTS ARE USUALLY INEFFECTIVE

Government-sponsored R&D is important and has had notable success when the government was the end user. The Manhattan Project, the Apollo program, the nuclear submarine, and aircraft engine programs are outstanding examples. We have been so impressed with our successes in these cases that we have come to believe that whenever we have a major problem, it can always be solved by a massive government program—technologically oriented and backed with unlimited financial resources. "If we can put a man on the moon, we should be able to do anything" is the refrain. What we do not understand is that in these success stories, the government was the final user of the products of R&D. In energy, it is different. The government is not the final user. This has serious ramifications: few government-sponsored energy research projects have borne fruit.

In the commercial sector, we are also learning that lesson. The Electric Power Research Institute is not the user of the research it funds. The adopters and users are manufacturers and utility companies. EPRI is learning a hard lesson. When it wanted to support an idea it thought would be useful, the manufacturers were eager to take its money. But when the time came to tool up for manufacturing for production, the criteria changed, the decision makers changed. Eager partners for R&D became reluctant brides. EPRI has worked hard to solve that problem. Cost-sharing partnerships are becoming popular. If there is no promise of a real market (a supplier and a buyer), there is little if any support at EPRI for a research program.

In concluding this chapter, we cannot help remembering a personal incident. When the Department of Energy (DOE) was formed, every major company in the United States geared up to get "its share" of the R&D funds. One of the corporate planners in GE felt very strongly that it would be a waste of company resources to divert its top talent to competing for DOE money. He felt so strongly that he wrote to the CEO. He almost got fired, except that his boss, one of the authors, shielded him. Now, fourteen years later, history has proved that he was right.

## SUMMARY

The lessons described above point to the essential characteristics of energy systems for the twenty-first century. These systems must emphasize attention to contingencies (i.e., robust solutions), safety, reliability, economics, and environmental issues. They must emphasize that cleaner may in fact be less expensive in both the long and the short run. They must embody a different relationship between business and government in the energy sector.

Sir Isaac Newton once described human knowledge as grains of sand on a gigantic beach. We hope these lessons may be yet another grain. We are not so presumptuous to think that with them we will be able to avoid all future blunders. But we do hope that we will not repeat any. If you kick a mule and the mule kicks back, the first time you can blame the mule. If it happens a second time, the blame is yours.

How do we apply these lessons to future energy system planning? We will deal with this subject in the next chapter, beginning with the criteria for future energy systems.

## NOTES

1. *Reserves* is defined as the quantity of petroleum that can be economically produced in the future, given known technologies and expected prices. Although oil companies have been accused of deliberately understating reserves as a means of concealing assets, one should understand the technological and economic uncertainties involved. All reserve figures are at best arrived at indirectly by use of both judgment and calculations. There is no way to measure reserves directly. Reserves always must be estimated. This estimate is—at any point in time—necessarily a function of both technology and price. Although price can go up or down, technology can only improve.

2. "The largest naval armada assembled since World War II now patrols the Persian Gulf to guarantee the flow of Middle Eastern oil to the West." (Durning, 1988, p. 26.)

3. Under that arrangement, the manufacturing associates learn from GE how to make the stationary part of the turbine. GE supplies the rotating part and guarantees the performance of the entire turbine to the MA. The MAs are free to compete against GE in the world market.

4. An analogy may be helpful. The radio waves are—in part—the total of all radio signals being transmitted by each radio station. The total of all radio signals appears to be "noise," until each frequency is considered separately and identified as having a discrete set of messages. Mechanically, breaking the wind into components along directions at 90 degrees to each other yields orthogonal components, since a change in, say, the easterly component will not affect the other components.

## REFERENCES

Ackoff, Russell L. 1979. "The Future of Operational Research Is Past." *Journal of the Operational Research Society* 30, no. 2, pp. 93–104.

Adelman, M. A. 1987. "Are We Heading Toward Another Energy Crisis?" Oil Policy Seminar of the Petroleum Research Foundation. Washington, DC, September 29.

Alcamo, J., P. Kauppi, M. Posch, and E. Runce. 1984. "Acid Rain in Europe: A Framework to Assist Decision Making." Working paper number WP-84-32. Laxenburg, Austria: International Institute for Applied Systems Analysis.

Brooks, Harvey. 1986. "The Typology of Surprises in Technology, Institutions and Development." In *Sustainable Development of the Biosphere* edited by W. C. Clark and R. R. Munn. Cambridge: Cambridge University Press, pp. 325–347.

Cerf, Christopher, and Victor Navasky. 1984. *The Experts Speak*. New York: Pantheon.

Clark, W. C. and R. R. Munn, eds. 1986. *Sustainable Development of the Biosphere*. Cambridge: Cambridge University Press.

Deming, W. Edward. 1982. *Quality, Productivity, and Competitive Position.* Cambridge, MA: MIT Center for Advanced Engineering Study.

Durning, Alan. 1988. "Setting Our Houses in Order." *World Watch* (May–June), p. 26.

El Masri, M. A. 1985. "On Thermodynamics of Gas Turbine Cycles." ASME *Transactions* 882/Vol. 107.

Hettelingh, J. P., and L. Hordijk. 1987. "Environmental Conflicts: The Case of Acid Rain in Europe." Research Report RR-87-9. Laxenburg, Austria: International Institute for Applied Systems Analysis.

International Institute for Applied Systems Analysis. 1985. Annual Report. Laxenburg, Austria.

Lee, Thomas H., John C. Fisher, and Timothy S. Yau. 1986. "Is Your R&D on Track?" *Harvard Business Review* (January–February), pp. 34–44.

———. 1988. "Technology and Implementation of Combined Cycle Systems." In *The Methane Age*, edited by Thomas H. Lee. Dordrecht/Boston/London: Kluwer Academic Publishers.

Lidsky, Lawrence, M. 1983. "The Trouble with Fusion." *Technology Review* 86 (October), no. 7, pp. 32–44.

Morgan, Chris, and David Langford. 1981. *Facts and Fallacies.* Exeter, England: Webb & Bower.

Song Jiang. 1988. "China's Development Policy in Science and Technology." Working paper number WP-88-35. Laxenburg, Austria: International Institute for Applied Systems Analysis.

Tabors, R. D., and D. P. Flagg. 1986. "Natural Gas Fired Combined Cycle Generators, Dominant Solutions in Capacity Planning." *IEEE Transactions on Power Systems* PWRS 1, no. 2, pp. 122–127.

Technology Assessment Group. 1982. *Development of Methodologies for Planning of Research and Development Programs, Phase II—Final Report.* Schenectady, New York. Submitted to the Electric Power Research Institute.

U.S. Government Task Force. 1970. *The Oil Import Question: A Report on the Relationship of Oil Imports to the National Security.* Prepared by the U.S. Cabinet Task Force on Oil Import Control; the task force included George Schultz, secretary of the treasury; William Rogers, secretary of state; Melvin Laird, secretary of defense; and Walter Hickel, secretary of the interior.

White, D. C., et al. 1981. "Strategic Planning for Electric Energy in the 1980s for New York City and Westchester County." Cambridge, MA: MIT Energy Laboratory.

# Part III
# Prospects

# 7 A New Paradigm: Integrated Energy Systems

*Anything that is said that is new is not true, and anything that is said that is true is not new*—Anonymous

## INTRODUCTION

The previous chapters point toward a set of characteristics that will be required for energy systems of the next few decades. The path in energy systems development has been, to date, far less than straight. There have been, however, lessons learned along the way that, we hope, will stand us in good stead for the future.

Just as the lessons learned are obvious once they are pointed out, so are the prospects. Future energy systems are, a priori, neither more complex nor simpler than those of the past. They will be different. Concerns with robustness, flexibility, reliability, security of supply, and environmental cleanliness will characterize the energy systems needed to get us into the twenty-first century.

Dealing with these concerns and maintaining economic viability will require that the systems take advantage of greater potential for integration. Much as a steel mill or an oil refinery is, from the perspective of energy as well as the balance of materials, an integrated system, so will our energy and environmental systems need to be more fully integrated.

This chapter presents a set of snapshots of what we believe will be the structure of energy systems over the next few decades. They are referred to as Integrated Energy Systems (IES). On the surface they seem novel from the perspective of current practice because they do not fit into the neat pigeonholes afforded by current industrial structure. They are horizontally rather than vertically integrated. They involve multiple inputs and multiple outputs and, frequently, an intermediate product. The systems do not defy economics. To the contrary, they improve economics. They do look to increased horizontal integration,[1] but not far beyond the examples that can be seen today in other industrial sectors, and within the vision of what has been occurring, for example, in the electric power industry in the United States.

We tend to forget that such institutions as the family farm were far more integrated in the past century than they are now. The basic inputs of land, labor, seed, and other factors produced all of the goods for the farm and market. There were virtually no residuals in the system. Scraps from the table went to the pigs and the manure of these and other animals became fertilizer. There was little if any waste in the system. Energy, like all other products in the system, was fully exploited.

Chapter 6 pointed out what we believe are the lessons of the past twenty years. Chapter 7 paints a rational picture of the next twenty years, presenting one approach to developing energy systems, integrated energy systems, that more nearly fulfill the needs of society today and for the foreseeable future.

## The Critical Characteristics

In assessing today's evolving energy system, there is, as was discussed in the previous chapter, a set of critical characteristics that must be satisfied. Systems must be

Clean,

Secure and reliable,

Safe,

Economic, and, most significantly,

Robust.

Each characteristic is heavily influenced by evolving social demands and energy supply dynamics. All five require that we move gradually, from the vertically integrated systems of the past to more horizontally integrated ones.

**Cleanliness.** The environmental movement that began in the 1960s intensified in the 1980s and will continue to do so into the twenty-first century. Energy conversion and use are correctly perceived as major contributors to our environmental woes. Whether acid rain is caused in part by coal-burning power plants, or smog stems from automotive exhaust, or the greenhouse effect is caused in part by an increase in carbon dioxide from fossil fuel combustion, energy and environmental cleanliness are inextricably connected. In the past, we made two fundamental mistakes: we believed the biosphere could absorb whatever we generated; and we thought we could always find a technologic fix. We have now learned that the best way to deal with a problem is to prevent it; removal of sulfur from coal or oil prior to combustion is better (though it may initially seem more expensive) than installing scrubbers in the stack.

**Security and reliability.** No one likes surprises, including abrupt changes in energy price and availability. Over the years we have looked largely to single sources of supply to provide the bulk of our needs—wood, coal, or oil. Security of supply implies the ability to provide the energy for end use from more than a single source, whether that is a geographic dispersion of sources or a set of alternate fuels. When we think of security of supply we generally consider time measured in years, possibly decades. We mistakenly thought we had stable oil supplies and prices in the early 1970s because we were dealing with "friends." The world changed and we found that we had all of our eggs, at the margin of our supplies, in very few baskets, politically, geographically, logistically, and qualitatively (i.e., mostly petroleum). We learned to plan for contingencies—no matter how unlikely they may sometimes appear.

The term *reliability* has most often been used for shorter time frames, on the order of hours, weeks, or months. What is now becoming clearer is that in our industrialized societies reliability has become an essential component of the production process.[2] Unanticipated interruptions are simply not acceptable.[3] We have learned that security and reliability are a continuum and together they define part of the fifth characteristic, robustness. Energy systems must be capable of adapting to both the long and the short term regardless of the term used.

**Safety.** How safe is safe enough? One of our colleagues has even asked, "Is Idiot Proof Safe Enough?" (Bucciarelli, 1985). The answer changes over time, moving toward greater safety. Safer is preferred both in the societal sense of having fewer in the population at risk of exposure to the residuals of the energy system, and in the particular sense of minimizing risk to individual employees in facilities or individual users of an energy technology. The U.S. experience with civilian nuclear reactors is the case in point. The experts can prove that it is safer living next to a nuclear power plant than walking across the street. Yet it only requires a Three Mile Island or Chernobyl to highlight that such analogies are irrelevant even if they are true. Safe by the old criteria is not enough because the systems can fail. For a reactor to be acceptable in today's environment, for example, society needs to be convinced that a reactor that fails will fail safe; it will not just cool down but will emit no radiation.[4]

In general the demand for the safety of large systems may be thought of in terms of how failure occurs. Terms like *fail soft* are being used to capture the idea that a system, when it fails, should contain the damage of the failure within as small a unit as possible and not affect human life. One example is large-scale flywheels. These are now made of a wound fiber, kevlar, rather than solid metal. When they fail, the result resembles a fishing reel after a poor cast rather than a bomb exploding. Safety, like cleanliness, is best achieved by perceiving the requirement as an integral part of the design rather than as an add-on after the product has been designed.

**Economics.** Energy is not a commodity that we consume directly. It is a factor of production and therefore a derived demand. In addition, energy is only one factor of production and frequently a small one for which capital or labor, or both, can be substituted. Our mistake in the 1970s was thinking that no adjustments to demand as a function of price were possible. We believed that we had to replace—BTU for BTU—our imported oil or suffer a retreat in our standard of living. We now know this was wrong. Energy is really no different, in principle, from any other factor of production. The mix of the factors we use is a function of their relative prices. Furthermore, we now recognize that the ability to switch energy sources as prices change may be desirable and worth the investment in capital needed to make it possible. To many electric utilities this has been an operating principle for at least the past fifteen years; many maintain dual fuel capabilities between natural gas and oil wherever possible. Consolidated Edison of New York, a major gas and electric company, provides an excellent example of economic switching in which gas is used for electric generation during the off-peak summer months when there is no higher-value market in residential systems and when the price of gas is below that of oil.

**Robustness.** "The forecast is always wrong!" The late Prof. Fred C. Schweppe made this statement famous at MIT and in the electric utility industry. His meaning is critical here. As long as you make a single forecast based on one possible set of assumptions concerning the future, you will most likely be wrong, because the probability of a specific set of events occurring in the future is extremely low. We have looked at the blunders caused by inaccurate assumptions (energy demand is not price sensitive), by incorrect measurement tools (economies of scale hold forever), and by applying the wrong paradigm (government investments in technology can significantly influence the private marketplace). At the time the assumptions, measures, and paradigms did not seem bad. We now know that we need to go beyond our assumptions; we need to look for contingencies that can take our solution out of the range of the acceptable. In short, we learned that *the best answer is the one that, under a broad range of contingencies or scenarios about the future, provides a good solution.* It is the robust rather than the optimal alternative that is preferable.

Robustness, cleanliness, security, safety and reliability, and economy are the major characteristics demanded of any energy system by today's industrialized societies. As a set, they reflect our desire to move beyond a focus on a short-run optimal mix of goods and services to consideration of the quality of life for both present and future generations. These desires will, we believe, result in evolutionary changes in the structure of our energy systems. Society is seeking robust systems and is willing to pay for them.

## THE NEW PARADIGM: THE INTEGRATED ENERGY SYSTEM

The desires of society and the role played by energy in our economic system have changed since 1973. Several questions must be answered.

- Are there ways of evolving energy systems that fulfill the criteria described above?
- Is it possible to accommodate the new or expanded requirements at a cost that is acceptable?
- Is it possible to develop energy systems that will provide for today's needs and adapt to tomorrow's needs and technological advances?
- Is it possible to build this system with today's technologies or must we await a major technical breakthrough, such as the development of a new photovoltaic material or a workable thermonuclear fusion reactor?

The answers must be seen in a context of change in the way in which energy systems, and industrial systems, are perceived: a new paradigm. The old paradigm separated each primary source of energy supply and provided only for vertical integration (e.g., from the oil or gas well to the consumer's heater). The new paradigm provides for a horizontally integrated energy system whose structure could range from extensions of industrial cogeneration, as we are beginning to see it today, to more complex integration that might link emissions (e.g., carbon dioxide) from an electric power generator to enhanced oil recovery, or might link a walk-away safe nuclear generation facility to a steam reformer.[5]

In each case, integration brings the subsystems together within a framework that allows for greater efficiency of use, the use of both existing and emerging technologies, and the achievement of the goals of cleanliness, security and reliability, safety, improved economics, and robustness.

How might integration be accomplished?

Before answering this, we need to take—for a moment—a giant step forward to see what some of the system options, combinations, and permutations might look like. Then we can return to the industrial and institutional realities of today to see how one or more of the paths of integration might be achieved to meet the goals described above.

## The Giant Step Forward

The structure of an Integrated Energy System provides a new paradigm for energy systems. Even though the integrating structure is new, it is assembled from existing parts. Its basic assumption is an integration or connection of the input fuels to the output demands through a set of transformation processes that produce industrial gases such as hydrogen, oxygen, and carbon monoxide and use other gases such as methane.

Let us consider a simplified representation of an Integrated Energy System. (See Figure 7-1.) On the lefthand side of the figure are the basic inputs: solid, liquid, and gaseous fuels. In the middle are the carriers of the energy and chemical value of the fuels, generally industrial gases. On the righthand side are the final transformations of the fuels into their final products and points of consumption, whether thermal or chemical energy. The diagram indicates the linkages within the system, from initial fuels through final utilization.

In the new paradigm there are many structural *options* in delivering the services required of an energy system. Many of the options are significantly different from the pattern we see today, where a fuel is used for a single final purpose and the residuals are left in the environment. In the new paradigm, final demand may be supplied

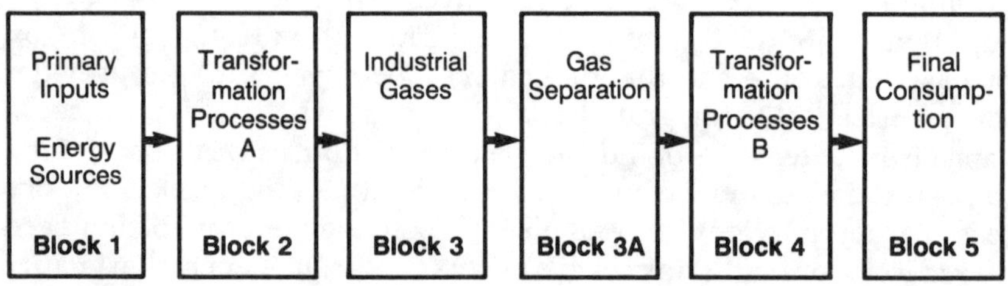


**FIGURE 7-1 Integrated Energy System: Simplified Representation.**

from more than one source and environmental residuals are explicitly a part of the structure, generally as a by-product of the process where they can be sold at a value or can be handled in a concentrated rather than a dispersed form. The new system also separates, in the intermediate stages, any single input source from a specific output product. There are many paths through the system to satisfy the end use. The existence of many options allows for a more robust system.

The IES structure combines the options in such a way as to improve economics, safety, and security, and reduce environmental damage. The diagram, then, is the framework for an interconnected structure and serves as the basis for further design and analysis. A more complete picture of IES (see Figure 7-2) depicts multiple paths for moving from a given set of resources to a required set of end energy forms. The choice of a path depends upon a variety of local factors: the availability and price of resources, the demand profile, pricing policy, capital stock already in place, and environmental concerns, to give a few examples. Once the paths are determined, the IES diagram forms the basis for synthesizing a set of robust and flexible options.

An additional and highly significant difference in the new paradigm is the relationship between production and control of environmental residuals. The IES system captures the residuals before they are emitted into the environment by turning them into by-products or capturing them immediately upon production. Rather than becoming pollutants in the biosphere, residuals are made into valuable products or captured as produced when they are not by-products.

**Integrated energy systems: a closer look.** Engineers, systems analysts, and even economists prefer a basic single-line diagram in describing the interrelationships between components of the system (see Figure 7-2). What differentiates this diagram from others is that even though there are a number of boxes representing technological steps, *the boxes represent only options—not required components.* The IES diagram offers alternatives in each stage of the system.

- Block 1 (Figure 7-1): Energy Sources, plus air and water . The block represents the raw materials that enter the system. Here, fuels begin to be *processed,* rather than only being consumed for their thermal value. This is an important difference from many if not most other energy systems.
- Block 2: Transformation Processes (A). Incoming fuels are transformed into industrial gases or liquids through energy conversion processes (such as steam reforming, gasification, and liquefaction). Air and water provide oxygen, hydrogen, and

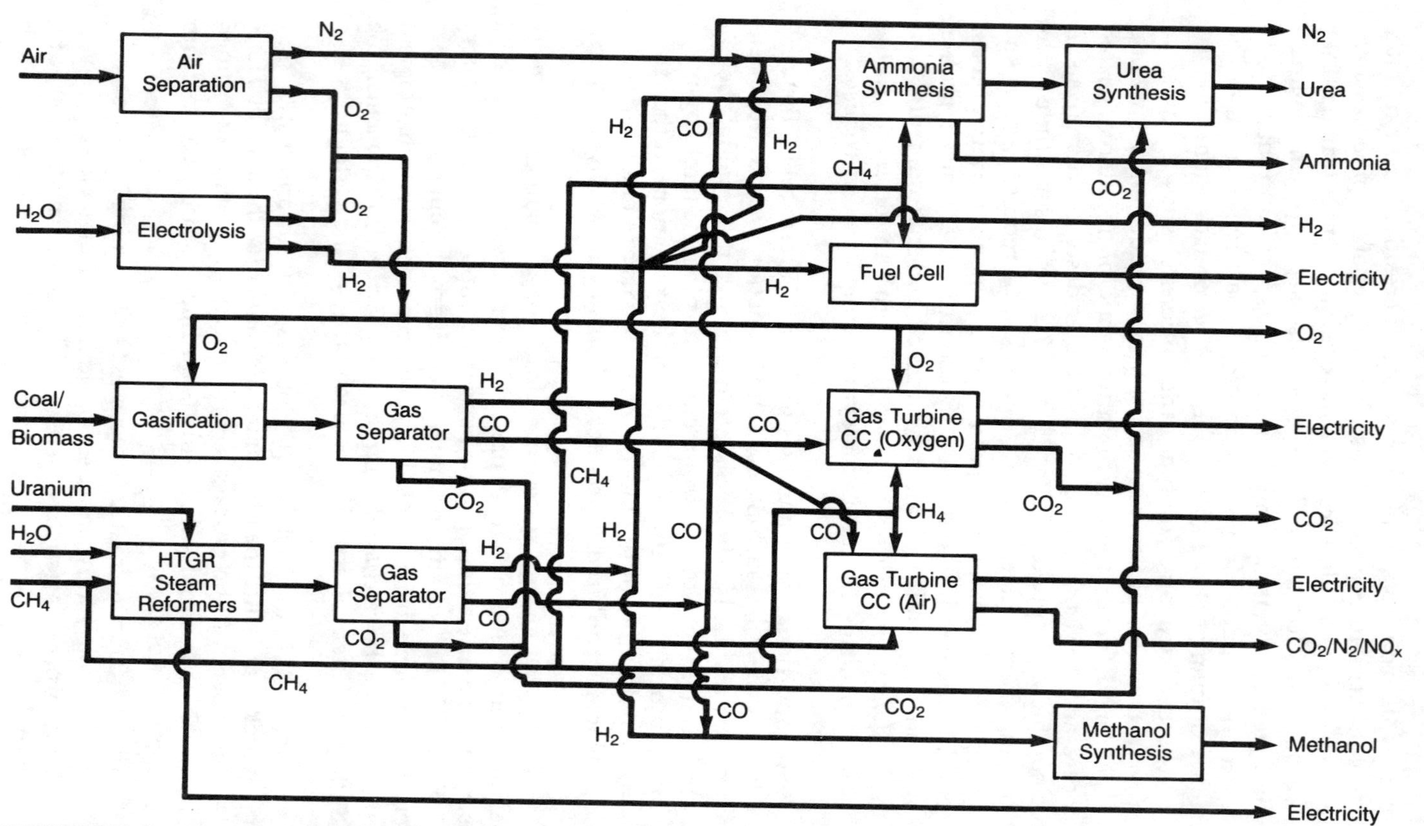


**FIGURE 7-2 Horizontally Integrated Energy System.**

nitrogen. At this stage, the thermal energy required for the transformation may be supplied by carbon-containing fuels, or by nuclear-based fuels (i.e., from an HTGR).[6]

- Block 3: Industrial Gases. A unique aspect of the IES is the manufacture of industrial gases. These include carbon monoxide, hydrogen, oxygen, and nitrogen, all of which have, or can have, commercial markets. In addition, natural gas or methane is carried forward as an industrial gas in the system. With the exception of methane, the gases are based on transformations that occur in Block 2. Also with the exception of methane, all gases can be generated from multiple processes.
- Block 3A: Gas Separation. Industrial gases are a critical component throughout the IES. As described in Block 2, the technology of air separation can be used to separate oxygen from nitrogen. Further downstream in the IES, the gas separation technology may be needed or may be economically desirable for generating hydrogen for a fuel cell or for use in ammonia synthesis in urea production.
- Block 4: Transformation Processes (B). The second transformation changes industrial gas into a more usable energy form, frequently electricity from combustion, or electrochemical reactions. To appreciate the robustness and flexibility of the IES, notc that, like the production of industrial gases in Block 2, the Block 4 technologies can receive inputs from more than one source.
- Block 5: Final Consumption. There can be multiple outputs from the IES, which can fulfill the same needs of society. These are multiple fuels, multiple chemical products, *and* multiple residuals. This advantage is one reason for IES robustness.

The IES structure offers sets of paths through the network that are neither all-encompassing nor exclusive. The following section examines two existing systems, two evolving systems, and one future system. What is important is that some are extremely simple, yet contain all the principles of the integrated system, including the advantages of flexibility, robustness, cleanliness, and choice. Other systems are more complex, yet still demonstrate these characteristics.

An IES may be best described as a system that acknowledges, accommodates, and exploits the interactions among major components in the energy system and the end use. The integration focuses on the relations among the initial source of energy—the fuel—its transformation into an intermediate product, and the use of that intermediate product in final demand as thermal, electrical, or chemical energy. Economics provides a context. That is, the economics must be favorable. But the economics applies in a system whose boundaries are appreciably larger than today's traditional definition.

The old energy paradigm separated types of energy use, and the manufacturing industry from the energy-supplying industries. The supplier of natural gas was, for instance, at least three transactions removed from the consumer. Gas separation, trunkline transportation, and local distribution stood between the supplier and final user of natural gas. In addition, the supply side is separated into a natural gas industry, oil industry, coal industry, and nuclear fuels industry. The new paradigm brings the industries together within a framework that allows for more efficient energy utilization and the use of existing technologies within the framework, and achieves the goals of increased environmental quality, security, and reliability. It allows for the existence of technologies or systems of technologies that can adapt over relatively short time periods to interruptions in supplies, and are capable of evolving as society's attitudes toward environmental quality and risk change. Critically too, the systems are sufficiently robust to accept new technologies when they are developed. An IES is not new, though some components may be. An IES does not require miraculous revolutions in technology development.

## A Small Step Back

Having taken a giant step forward, we can now see the way in which a small step from where we are today moves us forward within existing industrial and institutional structures. Integrated Energy Systems are not new. They are a new way of seeing the world—a new paradigm for energy systems. Such systems have always existed. Ask any industrial firm that generates both steam and electricity (an integrated system). Ask any crude oil refiner who produces fuels, consumes some, and generates its own steam power and electricity. Though the refiner considers the operations just good business, the system is an integrated one. Ask a steel mill operator who cokes coal to produce coal gas used for the production of both chemicals and fuel, and coke used for chemical reduction and fuel. This is how steel is made; it is also an integrated energy system.

Existing systems include cogenerators and combined cycle systems at the smaller end, and oil refineries at the larger end. Evolving systems include carbon dioxide recovery from combustion processes or from ammonia production, and the integrated gasifier combined cycle system developed by EPRI. Advanced systems include such evolving technologies as the high-efficiency gas turbine, combustion in oxygen rather than air environments, and developmental nuclear technologies such as the high-temperature gas-cooled reactor (HTGR). The IES as a whole reflects increased opportunity for interconnection between the elements, greater economic efficiency on a potentially smaller scale, and greater product availability, and with it greater choice.

On a first inspection, the IES might appear to be a complex and unmanageable system, akin to those highly centralized, oversized, and inefficient systems we have criticized throughout the book. Further, someone might argue that only governmental actions could bring them into existence, actions that we fervently argue should be avoided. In fact, evolutionary economics brings them into existence as it has simple systems like cogeneration and combined cycle generators—as long as the governmental and institutional frameworks of our society do not work against them.

The remainder of the chapter focuses on examples of Integrated Energy Systems that fulfill some or all of the goals described above. No system is correct, a priori. Many of the configurations described will be impractical in some or all locations. The critical point is that the paradigm be examined to see if the integration passes the market test *before* it is rejected as too costly or too complex.

**Commercially available integrated energy systems.** We can draw illustrations of existing integrated energy systems from two industries: petroleum and steel. A typical oil refinery and petrochemical complex and a typical steel mill are excellent examples of integrated energy systems that have worked well for decades, although these industries did not intend to build "integrated energy systems."

*The oil refinery and petrochemical complex.* The energy system that is a part of a typical large modern oil refinery and its associated petrochemical plant exhibits the features of an Integrated Energy System in a number of ways. (See Figure 7-3.)

There is no clear distinction between product streams and energy streams. Crude oil, liquefied petroleum gases, natural gas, and other industrial gases are the primary materials used by the complex, but each is used for many purposes. For example, natural gas may be used as fuel in heaters, or as feedstock in a unit making hydrogen, which will become part of a petrochemical product. In addition, natural gas may be used as fuel for the unit making hydrogen.

Industrial gases are exploited for their maximum benefit. One of the reforming units that upgrades gasoline might also produce hydrogen, which will become part of a petrochemical product, while using as fuel the gas made as a by-product from the cracking of residual fuel to gasoline. The entire steam system is integrated to exploit the advantages of operating at higher temperatures and

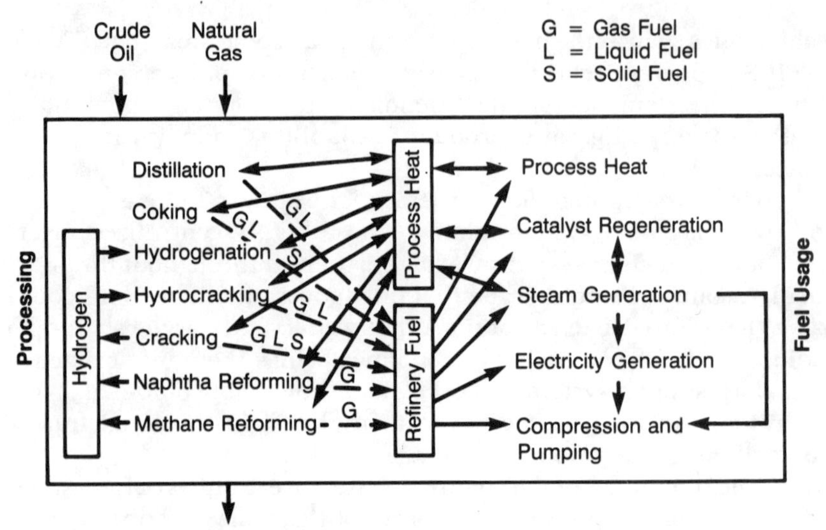


**FIGURE 7-3 Oil Refinery.**

thereby over greater changes in temperature. This relationship is known as taking advantage of the second law of thermodynamics.

There are a number of reasons why petroleum installations provide such good examples of integrated energy systems.

- There is every reason to make the system as efficient as possible; it is their nickel.
- The energy systems for the refinery and petrochemical plant were meant to be fully integrated into the total complex, not separable subsystems.
- There was no distinction between fuels and energy in any form and in either feedstocks or products. You cannot tell the difference between the inputs and outputs except on the bottom line.
- Industrial gases (hydrogen, oxygen, nitrogen, carbon monoxide, carbon dioxide, methane, ethane, ethylene, and so forth) are exploited for their greatest value, as fuel, raw material, energy carrier, or end product.

The result is a robust, flexible system, highly efficient in both energy and capital terms, and therefore economically sound. In other words, an Integrated Energy System. The environmental and safety issues are faced by eliminating problems rather than fixing them. Fuel and energy substitution are faced on a systemic basis.

The fact is, no one would ever think of building or operating a modern oil refinery and its associated petrochemical plants any other way, whether it was called an integrated energy system or not.

An important issue in the evolution of integrated energy systems is the removal of the informational[7] and institutional[8] barriers traditionally separating fuel sources, conversion systems, and ultimate users. Evolution toward appropriate Integrated Energy Systems requires overcoming two major obstacles. The first is our natural tendency to focus on only a part of the problem, to suboptimize.[9] The second is the lack of understanding of the real needs and possibilities of each element in the system.

The Nynashamn Energy Chemical Complex in Sweden (Johnsson, 1986, and Hook, 1986) has extended the petrochemical complex described above to include as a major component thermal energy for sale outside the facility. The Nynashamn facility will produce ammonia, fuel gas, hydrogen, sulfur, liquid oxygen, nitrogen, argon, and carbon dioxide and will provide what would normally be an unwanted residual—waste heat—for district heating to the southern half of Stockholm. The project is privately funded and will be based on the environmentally sound utilization of imported coal. As a privately funded project it is market driven: all the products and by-products are sold in the market.

*Steel mill.* The example of a conventional steel mill is even more spectacular. (See Figure 7-4.) The primary raw materials are coal and iron ore. Despite the mill's huge need for energy, *it does not burn the coal.* It uses it as a *chemical raw material.*

The steel mill can be considered a chemical plant that releases energy in the process of manufacturing chemicals, an exothermic chemical plant.

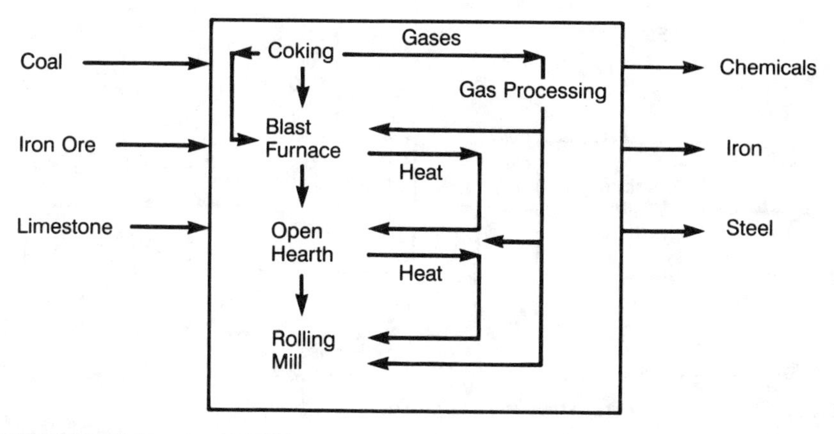


**FIGURE 7-4 Steel Mill.**

The coal is coked. Using gases released from itself as a fuel, the coal is heated to remove all chemicals. These are recovered, and include naphthalene, pyridine, ammonia, benzene, toluene, xylene, and coal tar. The associated fuel gas is used in both coke ovens and open hearth furnaces. The coke is used as the chemical-reducing agent for the iron oxide in the blast furnaces, as well as their fuel. One of the important products of coal is heat; nevertheless, coal is not burned. This is an example of a horizontally integrated system.

**Evolving integrated energy systems.** Whereas the oil refinery and the steel mill are existing systems that are relatively complex, several of the evolving systems are inherently less so, specifically the gas- or oil-fired combined cycle system, their extension to the integrated gasifier combined cycle system, and systems that recover the carbon dioxide produced.

The gas-turbine combined cycle system is a decades-old technology. (See Figure 7-5.) The system combines a topping cycle based on a gas turbine with a bottoming cycle in which the hot gases are passed through a heat exchanger to generate steam. The efficiency of the system is highly dependent upon the exit temperature of the gas turbine, which in turn is dependent upon the design of the turbine itself, including forced cooling and reheating between stages (El Masri, 1986). Today's GTCC systems can achieve efficiencies as high as 45% to 50%, which is spectacular when compared with 38% efficiencies in good coal-fired power plants.

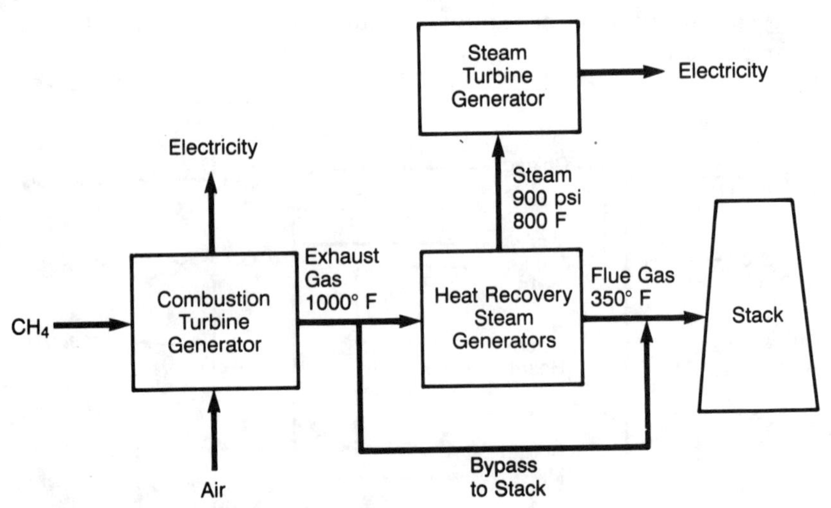


*Source:* EPRI. 1988. *Technology Assessment Guide*, Palo Alto, p. B-80.

**FIGURE 7-5 Natural Gas-fired Combined Cycle.**

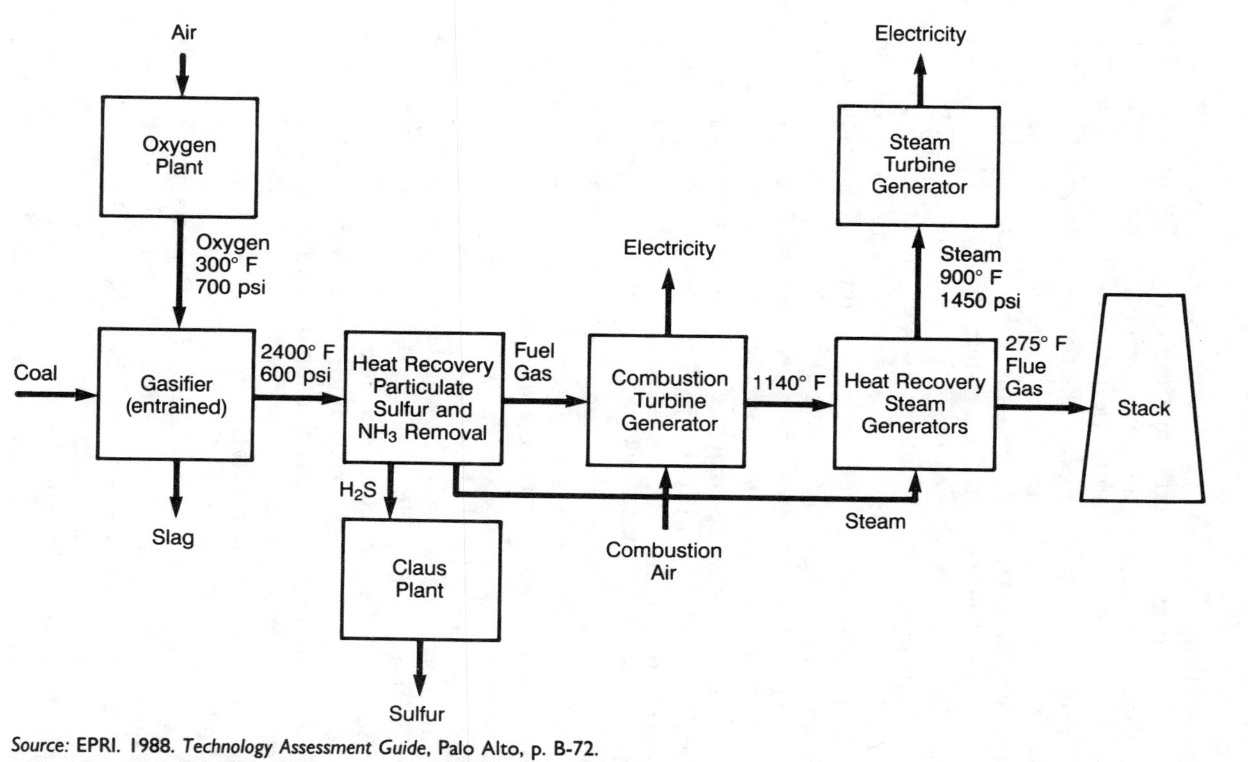


Source: EPRI. 1988. *Technology Assessment Guide*, Palo Alto, p. B-72.

**FIGURE 7-6 Texaco Integrated Gasifier—Combined Cycle.**

Development of the gas turbine has been driven by aircraft engine development, not by stationary applications. As discussed in Chapter 5, the result has been a system designed around weight and size rather than outlet temperature. Despite these constraints, GTCC systems have been used to some extent by the electric power industry and in cogeneration applications.

In the IES, the GTCC becomes a pivotal technology. The GTCC has the ability to utilize alternate fuels ranging from dirty liquid fuels (e.g., bunker fuel oil) to methane to carbon monoxide. In addition, it is possible to develop a turbine that uses oxygen from the IES, rather than air. The result is a highly flexible technology that is available in the marketplace today and can be developed to higher efficiencies (estimated to be in excess of 55% overall).[10]

Extension of the GTCC is already a reality (see Figure 7-6). The integrated gasifier combined cycle system (IGCC) has extended the GTCC one more step in the IES structure to include a packaged gasifier on the front end, adding multifuel capabilities and flexibility. As we discussed in Chapter 3, the IGCC is seen by many in the United States as a bridge technology for electric utilities, filling in the temporal and economic niche between the high capital cost, long lead time options of nuclear and coal. It is possible to get approval for a multi-staged project beginning with the construction of methane-fired gas turbines, followed soon by bottoming cycle steam turbines, and followed eventually by the (coal-fired) integrated gasifier.

Evolutionary systems can be extended one more step into the IES structure through the addition of a carbon dioxide recovery facility (see Figure 7-7). Carbon dioxide in relatively small quantities at very high levels of purity has a large number of economic applications; however, the market is limited. Scrubbing carbon dioxide from exhaust streams requires a far larger market and one that may be slightly less quality-conscious relative to price. Such a market exists in the old oil regions of the world, where carbon dioxide can be used for enhanced or tertiary recovery of oil. The first major, privately financed carbon dioxide stack-gas recovery process designed into a gas-fired combined cycle was operated in Lubbock, Texas, and fed the Garza, Texas, enhanced oil recovery (EOR) project. The exhaust gas was purchased over-the-fence by a ten-company consortium that owned the facility and the pipeline to Garza. Because of regulatory constraints, the electric utility was not part of the carbon dioxide project, but was the owner-operator of the generation facility.

Ellington and his coauthors used the capital and operating experience of the Lubbock plant to evaluate the potential for EOR from stack-gas scrubbing in the Oklahoma oil fields (Ellington et al., 1984). (See Figure 7-8.) The system economics is extremely sensitive

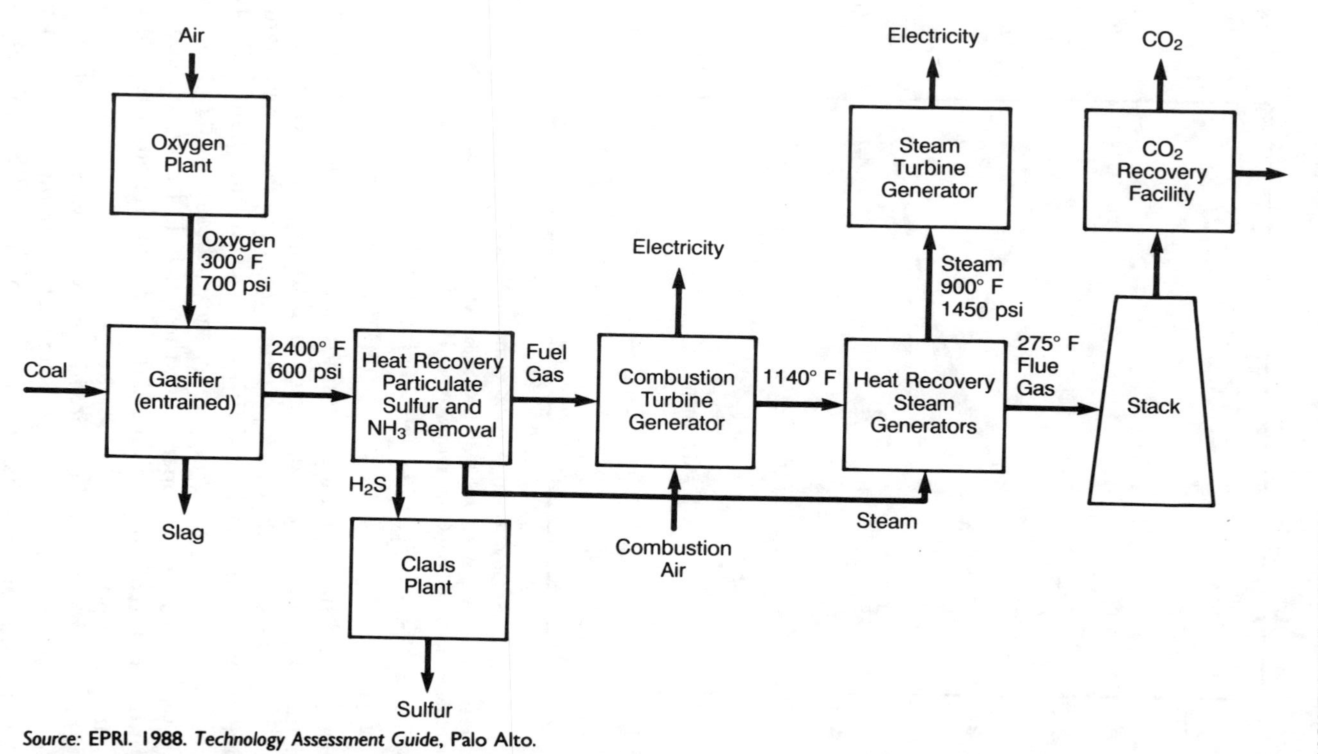


*Source:* EPRI. 1988. *Technology Assessment Guide*, Palo Alto.

**FIGURE 7-7 Texaco Integrated Gasifier—Combined Cycle with $CO_2$ Recovery.**

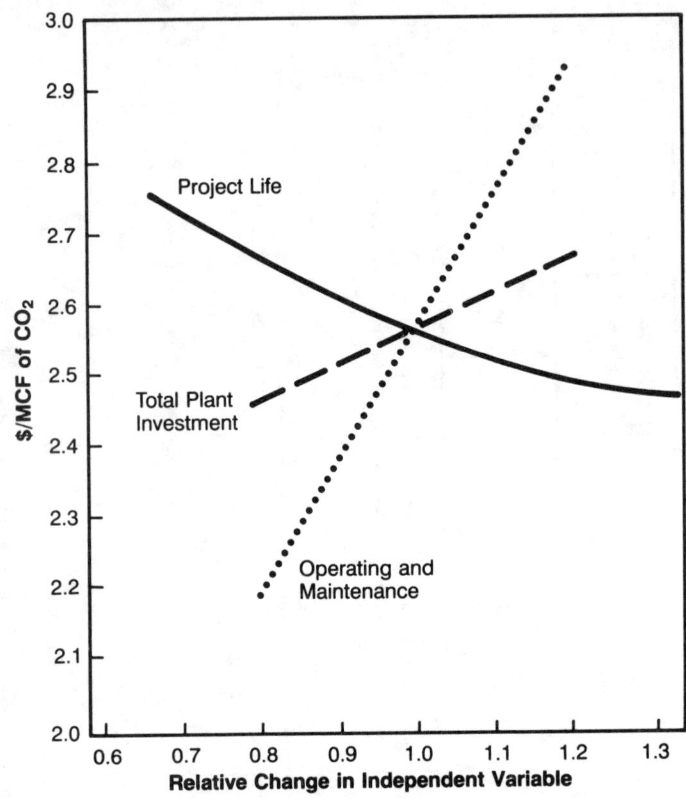


*Source:* R. T. Ellington, L. Warzel, B. Achilladelis, K. Saldanha, and M. J. Mueller. 1984. "Scrubbing $CO_2$ from Plant Exhausts," *Oil and Gas Journal*, October 15, p. 112.

**FIGURE 7-8 Sensitivity of Leveled Price of $CO_2$.**

to operating and maintenance cost and market oil prices. Given the cost of fuels and operations of the carbon dioxide facility, it is not difficult to understand that in its current form the facility is highly sensitive to world oil markets. With low-cost oil, the plant quickly becomes uneconomic. After one year, the carbon dioxide recovery facility was no longer in operation, though it is likely to become operational should oil move above $30/bbl, at which point the project will again be economic.

The fact that the carbon dioxide recovery system is not in operation at the time of writing may be seen by some as a negative point of IES. In fact, this is an example of how IES can provide increased flexibility in dealing with uncertainty in international markets. With prices low, the EOR oil is uneconomic and will not be recovered. When and if prices increase above the break-even level, the EOR becomes cost effective and the system again goes on-line.

This is a reasonably robust and flexible system for increasing production levels when required by either economics or social needs. As is almost always the case, *increased flexibility requires additional capital expenditure*, which may, at times, remain idle.

The Integrated Energy Systems described above are extensions of existing systems. Some are in place but uneconomic; others require additional R&D efforts. The following section takes the concepts one step further, first into an industrial complex and then into a fully linked system.

**New Integrated Energy Systems from existing technologies.** Industrial complexes, such as those along the Intracoastal Canal of the Texas-Louisiana Gulf Coast, provide a rich opportunity to exploit the IES concept. As described above, each petrochemical plant has long exploited—even if unknowingly—the IES approach, meaning that each plant operates as efficiently as possible. The whole complex is *not* optimal, however, in that the region could probably operate more efficiently and cleaner with greater integration.

The Texas-Louisiana Gulf Coast is known as the spaghetti bowl because of its multiple crisscrossing pipelines, which carry, from plant to plant, a wide range of charge stocks and intermediate and final petroleum and petrochemical products, including natural gas, propane, butane, isobutane, crude oil, refined petroleum fuels, ethylene, propylene, butylene, and naphtha. The pipelines connect plants lying hundreds of miles apart.

The ethylene pipeline system (now thirty-five years old) provides a classic example of an integrated energy system, applied to an industrial complex as a whole. Before the pipeline was built, ethylene—a primary ingredient in the petrochemical industry—was manufactured in each plant that needed it. In plants that made it as a by-product but had no use for it as such, it was considered no more than another component of plant fuel. The innovation was for one plant to manufacture and supply enough ethylene to a number of large plants to meet their needs, transporting it at high pressure and high purity through a pipeline a hundred miles long. Suddenly, high-purity ethylene became a utility, like drinking water. Now if a plant wants ethylene, all it has to do is tap into the line and install a meter. The next logical step would be for plants producing surplus or unwanted ethylene to meter *into* the pipeline grid, but this has occurred only to a limited extent. More important, the idea has not spread widely to other industrial gases: nitrogen, oxygen, carbon monoxide, carbon dioxide, and hydrogen. A pipeline carrying naturally occurring carbon dioxide from the Rocky Mountains into West Texas and Oklahoma for tertiary oil recovery has been built. But a plant requiring any of these materials must generally produce it on its own rather than take it from a utility of which it could be a part. In most cases, a plant does without the gases, because—on an individual basis—it cannot afford to produce them.

Another example of a new energy system based on existing technology is that of ammonia-methanol production. Both processes are based upon the input of a synthesis gas mixture (hydrogen and carbon monoxide). Traditionally the mixture has been produced by the reformation of hydrocarbons (most commonly methane). In the new integrated structure, there are many alternatives available for providing the required synthesis gas. In the generalized picture, streams of carbon monoxide and hydrogen become available from gasification processes as well as from the integrated HTGR/Steam Reformer/Steam Cycle (HTGR/SR/SC) module. The required gases are tapped off the common intermediated gas streams. The gases are free of potential pollutants; nitrous oxides and sulfur oxides are stripped off upstream.

In the case of ammonia, system integration can be carried one step further. Ammonia is commonly used in agriculture as urea, produced by combining ammonia and carbon dioxide. In conventional urea plants, the ammonia and urea synthesizing processes are integrated. The carbon dioxide stripped from the ammonia synthesis process is used in the manufacture of urea downstream. In the integrated system, the availability of an intermediate carbon dioxide stream makes the synthesis of urea a logical extension. At the same time, however, the availability of alternative uses for the carbon dioxide (in enhanced oil recovery, and so forth) ensures that the carbon dioxide input to the urea process is correctly valued.

Following the onset of the oil crisis, methanol emerged as a potential alternative to gasoline. Methanol has many advantages, primarily because it is a clean-burning fuel. Viewed in isolation, however, methanol production (traditionally based on natural gas) has not been proven economically viable. In the integrated system, where costs of production are allocated to include the environmental effects as well, this fuel *may* prove to be economically viable.

Major processing and manufacturing industries and the electric power industry have long understood the advantages of integrated systems. Dow Chemical in Midland, Michigan, generates electricity as a by-product of its thermal and chemical processes. General Electric and Big Three pipeline generate electricity for sale to Houston Power and Light. Nearly all of the older East Coast urban utilities (New York, Boston, Philadelphia, and Baltimore, to name only four) sold steam as a by-product of their electrical generating systems. In addition we are now seeing industrial gas pipelines. There is a major carbon dioxide pipeline in the southwestern United States and a hydrogen pipeline in Germany. We have a long history of integrated systems and transport of industrial gases; these will become more important as we enter the twenty-first century.

## THE ECONOMICS OF INTEGRATED ENERGY SYSTEMS

In the previous pages we have presented, in general terms, the structure of an Integrated Energy System. This section will take the general and make it more specific through a case analysis of the application of Integrated Energy Systems within the energy economy of the United States. Following a discussion of what we perceive to be the economic advantages of the IES, we will briefly look at policy and answer two questions:

- If Integrated Energy Systems are so good, why aren't they used today?
- What is the role of government if more integrated energy systems are to become a reality in the energy economy?

The International Consortium on Integrated Energy Systems, coordinated by the Massachusetts Institute of Technology, studied a detailed IES.[11] (See Figure 7-2.) The results, presented below, are based on estimates of the current costs of energy and capital in the United States. Other studies have been carried out for each nation within the consortium. Results from the Kernforschungsanlage (KFA) Laboratory have been published (Hafele, 1988), and others are anticipated in the future.[12]

What are the economic advantages of the Integrated Energy System over the standard, stand-alone systems? What technologies are attractive and why? Technologies that are part of the IES, and their capital costs are listed in Table 7-1. It is important to point out that the information is based on data received from manufacturers or suppliers of equipment in the United States. For technologies that have yet to see a market, we have used the best estimates available from developers and experts. The discussion that follows must be used with care, given the discussion in Chapter 5 on the dangers of using the cost of electricity as the numeraire in the analysis. Nevertheless, we believe that the information presented is sufficiently succinct to allow the reader to evaluate our assumptions and to alter the capital and operating assumptions to meet his or her understanding of the economic conditions.

The studies completed have used electrical energy as the com-

**TABLE 7-1 The Technologies of the Integrated Energy System**

| *Technology* | *Function* | *Size* | *Capital Cost* |
|---|---|---|---|
| Air Separation | Produce Oxygen | 6,000 cu. ft/day | \$24 Million |
| Gas Separation | Derive CO and $H_2$ | 60,000 cu. ft/day $H_2$ | \$63 Million |
| Gas-Turbine Combined Cycle | Generate Electricity | 70 megawatts | \$650/kW Electric |
| High-Temperature Gas-Cooled Reactor | Heat for Reformer and Electricity | 95 megawatts | \$1,500/kW Electric |
| Other Gas Handling Equipment | | | \$13 Million |

mon denominator for the sake of convenience and general application. The results could have as easily been expressed in terms of tons of coal equivalent, or barrels of oil, or BTUs. Throughout the text we have used kilowatt-hour as a unit of comparison and continue to do so here. The critical jumping-off point is what the cost of electricity would be if current technologies were used. We have calculated the cost of electricity from a range of electrical generating technologies. (See Table 7-2.) It should be noted, again, that the calculations are based on assumptions about the initial cost of capital, the cost of the input fuel, the efficiency of conversion, the number of hours the plant is on line and generating in a specific year, and, critically, the financial value of money (i.e., the discount rate).

In order to compare the potential benefits of an IES with the stand-alone structure of only electrical power generation, we have constructed a hypothetical IES that includes an HTGR, primarily for thermal energy, a steam reformer for conversion of natural gas to hydrogen and carbon monoxide, a gas-turbine combined cycle that we have designed on paper to utilize carbon monoxide and oxygen, and a system for gas separation that allows us to recover the carbon dioxide as a useful by-product. The system is designed to have as inputs nuclear fuels, natural gas, and air, and have as outputs electrical energy, hydrogen, and carbon dioxide. The economic evaluation of the system is based on standard utility calculations for the efficiency of the components and the handling of the capital costs. In our example we have *assumed* (this is a dangerous word) that there is a market for both the hydrogen[13] and carbon dioxide and that the benefits derived from the sale of these by-products will offset some of the costs that are attributed to the generation of electricity. In the calculations, we have assumed in our base case that the system will operate with 75% overall availability and that the combined cycle system will operate at 47.5% efficiency, a level that is achievable with today's

**TABLE 7-2 Costs of Electricity from Conventional Generation Technologies**

| Technology | Size MW | Capital 1988 $/kW | Fixed Charge Rate | Annual Cost $\$ \times 10^6$ | Heat Rate BTU/kWh | Life-time Years | Capac-ity Factor | Fuel Cost $/mmBTU Conventional | Capital Cost $/kWh | Fuel Cost $/kWh | O&M Cost $/kWh | Total Cost $/kWh |
|---|---|---|---|---|---|---|---|---|---|---|---|---|
| Nuclear Conventional | 1,100 | 2,880 | 0.12 | 380.16 | 10,530 | 30 | 0.65 | 0.73 | 0.061 | 0.01 | 0.01 | 0.082 |
| Nuclear HTGR* | 95 | 1,500 | 0.12 | 17.10 | 10,000 | 30 | 0.65 | 0.73 | 0.032 | 0.01 | 0.01 | 0.052 |
| Coal Conventional | 1,000 | 1,170 | 0.17 | 198.90 | 9,460 | 30 | 0.75 | 1.60 | 0.030 | 0.02 | 0.01 | 0.052 |
| Oil Thermal | 1,000 | 800 | 0.17 | 136.00 | 9,400 | 30 | 0.50 | 3.00 | 0.031 | 0.03 | 0.00 | 0.064 |
| Gas Thermal | 1,000 | 800 | 0.17 | 136.00 | 9,400 | 30 | 0.50 | 3.50 | 0.031 | 0.03 | 0.00 | 0.069 |
| Gas-Turbine Combined Cycle | 220 | 514 | 0.17 | 19.22 | 8,400 | 30 | 0.75 | 3.50 | 0.013 | 0.03 | 0.00 | 0.045 |

*Estimates provided by Professor Lawrence Lidsky, MIT, April 1989.

*Source:* Fuel cost data derived from MIT working group on New England Energy Systems and capital cost data from the EPRI, 1986, *Technology Assessment Guide*, Vol. 1.

technology. The calculations are based on the assumption that the system will last for thirty-five years.

It is always wise to ask, "How good are these assumptions?" The answer is that they are within the range of reason and, as will be seen below, the application of thoughtful sensitivity studies indicates that the results do not vary dramatically with changes in the assumptions. In the base case described above, the cost of electricity over the planning horizon being evaluated is roughly $.075/kWh, levelized, that is, financially averaged over the lifetime of the facility.

The IES is competitive with the defined alternatives (see Table 7-2). *Given the assumptions about markets for hydrogen and carbon dioxide,* the IES is slightly more expensive per kWh than all but the conventional nuclear systems. The story is not, however, over at this point. We have analyzed the sensitivities of the final output cost of electricity to *realistic* changes in the input variables. (See Figure 7-9.) Specifically we looked at the sensitivity to operation and maintenance costs (O&M), carbon dioxide by-product value, hydrogen by-product value, capital cost of the HTGR and steam reformer subsystems, efficiency of the GTCC, capital cost of the GTCC, and the price of natural gas.

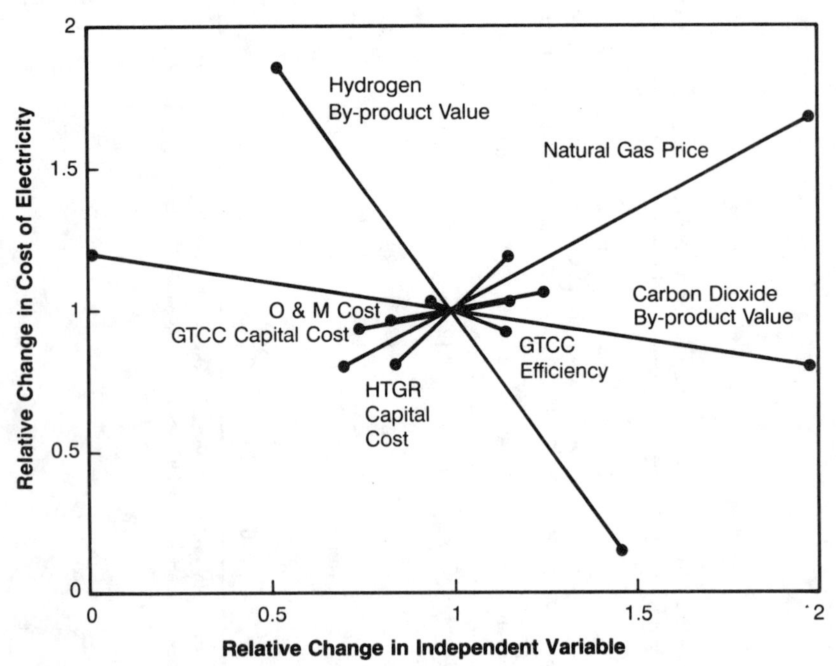


**FIGURE 7-9 Integrated Energy Systems Technologies Sensitivity Analyses.**

The Y axis of the figure presents the change in cost/kWh of the electricity generated normalized so that the base case ($.075) is set to 1. The X axis shows the change in the independent variable, the variable on which the sensitivity analysis is being applied. By presenting the information in this manner it is possible to plot the sensitivity of each parameter on a single graph. The sharper the slope of the line, the more sensitive the final price of electricity is to proportional changes in the independent variables. The length of the line reflects the amount of variation from the base value that we believed to be realistic; thus the change in efficiency in the GTCC could range only over a few percentage points while the by-product value for carbon dioxide could range by a factor of 2.

The results (see Figure 7-9) are clear. The output is very sensitive to assumptions concerning the market value of hydrogen,[14] and far less sensitive to the market value of carbon dioxide. This finding is significant because there is limited information about a market for hydrogen. In today's economy hydrogen is produced from the reforming of methane, an energy-intensive and expensive process that represents an upper limit on the value of hydrogen of roughly $3.15/mcf. The quantities being produced in our example of an IES are 50% larger than what today would be a large hydrogen production facility. It is likely, however, that an expanded market would exist at a lower price because hydrogen has value as both a chemical and a clean fuel. It is our belief that at lower costs many industrial and high-value fuel markets would open up. In our analyses, even fuel cells begin to look attractive with a lower cost of hydrogen.[15] The question is how much lower. Hydrogen's fuel value of 320 BTU/cu.ft. or roughly $1.15/mmBTU (with methane at $3.50/mmBTU) provides the lower limit. As a base case in the analysis reported here we have assumed an intermediate value of $2.15mcf. As can be seen from Figure 7-9, reducing the by-product value of hydrogen to its fuel value would eliminate consideration of this version of an IES, since it would be uneconomic. On the other hand, were hydrogen to have the market value of today's produced hydrogen ($3.15/mcf), the electricity (which would be an economic by-product) would have a cost of less than $.04/kWh.

The market price of carbon dioxide is no more firmly established than that of hydrogen. When oil prices are high, there is a market value for carbon dioxide for enhanced oil recovery. The market for manufactured as opposed to naturally occurring carbon dioxide disappears below about $30/bbl. The role of carbon dioxide as a greenhouse gas has, however, raised the question of whether the capture value may be nonzero, that is, whether there will be, in the short to intermediate run, a value attributed to being able to capture carbon dioxide rather than emit it into the atmosphere. Given the uncertainties, we have defined the by-product value of carbon dioxide by $0.00/mcf on the low end and $2.00/mcf on the high end. Again, it is clear that reducing the by-product value of carbon dioxide to zero increases the cost of electricity from $.075 to $.09/kWh.[16] This is a significant change, but not one that would eliminate the economic and certainly not the environmental attractiveness of this configuration of an IES.

Let us next consider the sensitivity of the results to other factors such as the capital cost of the HTGR. Of the capital costs, the cost of the HTGR is clearly the largest, and the one to which the system is most sensitive. On the basis of Lidsky's best estimates, the HTGR would show a cost of $1,000/kw (electric) or lower. We have used $1,500/kw as our base case and have taken a range of $1,000 to $2,000/kw for sensitivity analysis. This gives a range in the cost of electricity from $.065 to $.086.

What if, in reality, this configuration of an IES is not cost effective, as we have described it? The answer is that the *system* could be reconfigured (i.e., redesigned) to include different elements, or, if already built, could be evolved to include new elements. (See the appendix for a series of analyses using a linear programming approach for system design. The analyses show clearly the flexibility of the IES structure in identifying alternate pathways as a function of the cost of inputs and the value of the output products. The analyses point to an additional strength—the ability of the system to absorb new technologies.) Again, the question might be asked, what if a technology such as cold fusion proves successful? What will happen next? Given the structure of the IES, the answer is that it will be phased into a system that can be reoptimized around a new mix of technologies, one that produces electricity from a different mix of technologies and with a different mix of fuel inputs (deuterium, for instance) and an improved mix of environmental residuals.

In comparing technologies, we all too frequently compare the sensitivity of the lesser-known technology to a fixed value for the output of a better-known technology. This is, in fact, incorrect because the so-called known technology is also fraught with uncertainty as well, even if the band of uncertainty is less. To demonstrate this, we have used the same technique to examine the sensitivity of cost of electricity to uncertainties in the cost of both capital and fuel. (See Figure 7-10.) The band of reasonable uncertainty is narrower, but the impact of the uncertainty on the final outcome is clear. The technologies about which we supposedly know a great deal are just as, if not more, subject to the effects of future uncertainties in capital and fuel costs as are those of the IES, about which we know less.

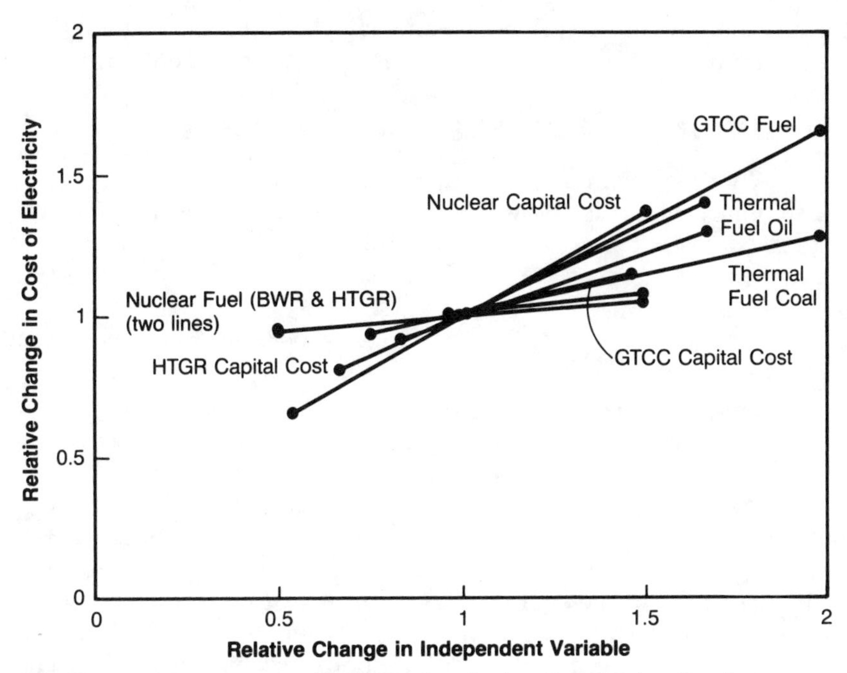


**FIGURE 7-10 Conventional Technologies Sensitivity Analyses.**

How well does an IES fulfill the criteria of robustness, flexibility, reliability, economic value, security of supply, and environmental cleanliness? As described above, IES is robust, because within a reasonable range of uncertainties the price of electricity we used as a numeraire is as stable as the same numeraire used for more conventional technologies. In addition, the system is robust and highly flexible with regard to changes in input and output values (see the appendix). When the value of the products and by-products changes, the system can easily be reoptimized to meet the new challenge. Reliability cannot be tested independent of operations. Suffice it to say at this point that reliability is generally a function of the size and complexity of systems and subsystems. Although the IES as a whole is a complex system, its components are small and relatively simple. Therefore, it is not technically difficult to design the system for given targets of reliability, although the design will have an impact on economics. And economics is essential for the system to operate. The two final criteria, security and environmental quality, are the real strengths of the IES. The IES has a high level of security because the system is flexible in its inputs and is environmentally sound because the residuals generated are costed and valued within the system; they are *explicitly not* handled as externalities left to the next generation.

## INSTITUTIONAL AND POLITICAL ISSUES

When we began this section, we asked two additional questions. These reflect the bridge between a good engineering idea and its final implementation: the institutional or political environment into which it must fit.

- If Integrated Energy Systems are so good, why aren't they used today?
- What is the role of government if more integrated energy systems are to become a reality in the energy economy?

The answer to the first question is easy. Integrated Energy Systems are not in place today largely because our current economic and business system is not horizontally integrated, but instead vertically integrated in the energy and chemical sector. This is not to say that such systems are not seen in today's economy. As we pointed out, an oil refinery and a steel mill are both examples of an IES. The Nynashamn Complex in Sweden is yet another example of an IES under development. Our energy and environmental concerns are at present undergoing a radical change. Global warming, ozone depletion, and acid rain (now an old issue) are forcing us to rethink the way in which energy is used in both developed *and* developing nations. In the economists' terms, there are now discussions on internalizing the externalities. Some of the social costs discussed in Chapter 5 may well become the responsibilities of the polluters in the future! We and others have developed Integrated Energy Systems to provide a new paradigm through which to look for technical and economic solutions to these problems.

Legal, informational, and institutional barriers do exist. We believe the move toward IES can be facilitated by recognizing the barriers for what they are, and working to lower them. As the reader should see by now, we perceive no need for government intervention or incentive. The private sector should require neither coercion nor bribes to do what makes economic sense. For this reason the second question is really the other side of the first one. Probably the most critical role of the government of the United States and other nations is to ensure that viable horizontally integrated alternatives will not be overlooked because of institutional barriers in the process of project development. This does not mean that the IES will be cost effective under all circumstances, only that the governmental and regulatory community, along with the business community, needs to be aware of the advantages to be gained from putting the pieces together in a new pattern.

## INTEGRATED ENERGY SYSTEMS: A SUMMARY

What are the advantages of an Integrated Energy System?

- *Clean:* An IES is based on the premise that it is desirable and possible to design cleanliness into the system as an integral part of the production process as opposed to taking the residual out at the back end.
- *Secure and Reliable:* The IES provides for flexible systems. Different societies and different environments require different systems. The IES structure contains both a series of technologies that are separable and recombinable and a set of fuels that can be substituted between technologies in either a short or a long term.
- *Safe:* The IES can be designed around safety from both a societal and a personal perspective. This safety extends to the nuclear and the nonnuclear technologies.
- *Economics:* The underlying assumption of the IES is that economic considerations are a dominant force in system design.
- *Robust:* Choosing a single system from among a number of options is only the first step. The more difficult and more important step is choosing a system that can evolve as technology, energy costs, and societal needs change.

Throughout the 1970s and early 1980s we sought to identify technologies that were then no more than a glimmer in the eye of the inventor or developer. We expected them to provide us with immediate solutions, BTUs and kWhs. IES offers a different approach. The IES takes existing technologies and looks for better ways to put the pieces together. The IES seeks marginal improvements in performance of components that can lead to major improvements in system efficiency or social acceptability, or both. Because technology is undergoing constant development, its evolution must be planned, augmented, and exploited. IES can absorb new technologies as they arise.

Given these characteristics, Integrated Energy Systems respond to the uncertainties of the energy system as we have defined them, uncertainties in supply, technology, environmental regulations, and social requirements.

Because of these advantages, Integrated Energy Systems are now under study in Japan, Sweden, Germany, Eastern Europe, and the Soviet Union. IES holds the promise of increased economic efficiency, lower environmental damage, and greater responsiveness to societal needs. It provides the robustness required for technological adaptability in the twenty-first century.

## NOTES

1. Today's energy industry is vertically integrated in large part. The energy companies frequently own from the source of supply through the delivery system, i.e., from the oil well to the gasoline pump. Horizontal integration implies that at the intermediate and the final demand levels the sources of supply may be owned or coordinated by the same company; that is, gasified coal, natural gas, and possibly hydrogen could be traded off at the margin and the final demand for heat or electricity or gasoline at the pump could be derived from any of the IES source fuels.

2. This does not mean that you have to have energy available all the time at a specified price, only that the energy needs to be there at some price. The work of Schweppe, Caramanis, Tabors, and Bohn (1988) bears witness that reliability is in the eye of the beholder, and, like other characteristics, it has a market value. You want it when you need it, and you need it when its marginal value exceeds its marginal cost.

3. It is important to note that the U.S. electric power and telephone systems, for example, are probably the most reliable in the world.

4. An example of a fail-safe reactor is a high-temperature gas-cooled reactor (HTGR) such as that developed by the Kernforschungsanlage Laboratory in Juelich, West Germany. Such a reactor is cooled by natural convection even in the event of circulating system failure (Hansen et al., 1989).

5. Steam reforming is a process by which a hydrocarbon, generally either a light oil or methane, is heated with steam to form carbon monoxide and hydrogen gas.

6. A high-temperature gas reactor operates at 900° to 1000° C as opposed to a conventional BWR or PWR operating at 300° to 350° C. This technology is included within the IES because it is considered walk-away safe and because it allows for continuous refueling, thereby improving the capacity factor of the plant relative to technologies requiring complete shutdown for plant refueling roughly every fifteen months. In addition, its higher operating temperature makes it suitable for use as a direct source of industrial heat, for example, to supply the endothermic heat of reaction in a chemical process. The possibility exists for economic, industrial-size reactors in the order of 200 MW (see Hansen et al., 1989).

7. Informational in that decision makers need to be aware of the potential for systems integration and the advantages from looking at a broader range of options.

8. It is critical to note that our legal and regulatory systems inhibit

integration in the energy area. Jurisdictional disputes in the regulatory area are well known. Legislative prerogatives are coming more to the fore and each primary actor is seeking different objectives and promoting different solutions to the same problem.

9. An established tenet of a market economy is that suboptimization by individual firms at market prices optimizes the market as a whole. Although this is true for the economy as a whole, it is not necessarily true at specific sites, for two reasons. The first is the institutional barriers to information flows and cooperation. The second is the problem of capturing fixed costs in a way that is appropriate for all parties.

10. Caution must be stressed at this point. Although it is possible to design a recycle system for carbon dioxide to mix with the incoming fuel and the carbon dioxide, the resulting cooling of rejected carbon dioxide is significant and will require research and development effort. The economics and feasibility of such an investment will depend heavily upon environmental constraints. Such a system would emit neither sulfur nor nitrogen oxides. Furthermore, the carbon dioxide could be a marketable by-product.

11. MIT's Laboratory for Electromagnetic and Electronic Systems and the Kernforschungsanlage (KFA) Laboratory of Juelich, West Germany, led the consortium. MIT coordinated the overall research effort. The other members were the Japan Atomic Energy Research Agency, a utility and governmental group in Taiwan, the University of Oklahoma, and the Institute for Hydrogen Studies, Toronto, Canada. The TRW Foundation provided MIT with resources for travel and writing. Air Products Corporation provided invaluable assistance to the consortium with its knowledge of the processes for air and gas separation, which are key to making the concepts work. The authors have served as director and principal investigator of the consortium since its founding.

The International Institute for Applied Systems Analysis in Laxenburg, Austria, assumed much of the role of coordination in the late 1980s. A group of Swedish private and public institutions was actively involved in the meetings and work of the consortium, though not formal members. Under the IIASA umbrella, the IES consortium was able to expand to include a number of Eastern European countries, specifically Bulgaria, an early participant in the MIT-coordinated consortium, the Soviet Union, and Czechoslovakia.

12. It has always been the plan of the IES consortium to complete a detailed monograph reporting the experience of the individual countries in their local studies of Integrated Energy Systems.

13. There is a very limited market today for both carbon dioxide and hydrogen. Carbon dioxide is the bubble in soft drinks (a high-purity, high-value product) and is used in enhanced oil recovery (a low-purity, low-value application). Hydrogen is more difficult to peg. At its high-value end it is used for hydrocracking in refineries and for chemical hydrogen in other processes. At its lowest-value end it is a fuel with a BTU value of roughly one-third that of methane, i.e., 320 BTUs per cubic foot.

14. It is important to note that the IES defined is almost a chemical plant in its current configuration. More hydrogen is produced than electricity on a value basis.

15. MIT carried out an evaluation of the economics of fuel cells in an IES (Malloy, 1985).

16. With carbon dioxide at $0/mcf, the cost of electricity would increase by a factor of 1.7 (see Figure 7-8). Therefore, $.06 (base cost) times 1.7 = $.10/kWh.

## REFERENCES

Bucciarelli, Lawrence. 1985. "Is Idiot Proof Safe Enough?" *The International Journal of Applied Philosophy* 2, no. 4, pp. 49–57.

Ellington, R. T., L. Warzel, B. Achilladelis, K. Saldanha, and M. J. Mueller. 1984. "Scrubbing $CO_2$ from Plant Exhaust Provides Economic Sources of Gas For EOR Projects." *Oil and Gas Journal*, October 15, pp. 112–124.

El Masri, Maher. 1985, 1986. "On the Thermodynamics of Gas Turbine Cycles." Part I, ASME *Transactions*, Vol. 107, October 1985; Part II, *Journal of Engineering for Gas Turbines and Power*, Vol. 108, January 1986; Part III, ASME *Transactions*, Vol. 108, January 1986.

EPRI. 1986. "TAG—Technical Assessment Guide, Vol. 1: Electricity Supply." Palo Alto, December.

Hafele, Wolf, H. Barnert, S. Messner, M. Strubegger, and J. Anderer. 1986. "Novel Integrated Energy Systems: The Case of Zero Emissions." In *Sustainable Development of the Biosphere*, edited by W. C. Clark and R. R. Munn. Cambridge: Cambridge University Press, pp. 171–192.

Hansen, Kent, Dietmar Winje, Eric Beckjord, Elias P. Gyftopoulos, Michael Golay, and Richard Lester. 1989. "Making Nuclear Power Work: Lessons from around the World." *Technology Review* 92 (February–March).

Hook, Anders. 1986. "The Energy System in Stockholm Today and Possibilities for the Future." Presented at the International Integrated Energy Systems Consortium meetings, Talberg, Sweden, September.

Johnsson, Stefan. 1986. "Status of the Nynashamn Energy Complex." Presented at the International Integrated Energy Systems Consortium meetings, Talberg, Sweden, September.

Malloy, John R. 1985. "System Design and Economic Evaluation of an $H_2$ Air Phosphoric Acid Fuel Cell Power Plant for All Electric and Cogeneration Applications." Master's thesis, Department of Mechanical Engineering, MIT.

Schweppe, Fred C., Michael C. Caramanis, Richard D. Tabors, and Roger Bohn. 1988. *Electricity Spot Pricing*. Dordrecht/Boston/London: Kluwer Academic Publishers.

# 8 The Future: One Approach, Many Paths

Our colleague, the late Ed Schmidt, used to say, "Energy policy is like defense policy. It varies from country to country, guided by political, economic, societal, and cultural considerations." To illustrate his point, he often used the striking example of nuclear power. The United States depends on nuclear weapons for its defense policy, but U.S. society largely rejects the nuclear generation of electricity. Japan is exactly the opposite. It rejects nuclear weapons for defense, but relies heavily on the nuclear generation of electricity. France, on the other hand, relies on nuclear power for both defense and the generation of electricity.

We are keenly aware that there is no single correct solution for an energy policy. Nor have we attempted to identify one. What we tried to do was identify a set of criteria that are truly generic, and applicable whatever the particular political, economic, and societal conditions may be in a particular location at a particular time. The five criteria we proposed reflect that effort. An energy system should be clean, secure and reliable, safe, economic, and robust.

In forming energy policy, the United States needs to pay particular attention to its own political, economic, and societal conditions, since its uniqueness is largely unappreciated. Among the facts to keep in mind are:

- The United States is both the largest producer and the largest consumer of energy.
- Along with Canada and the United Kingdom, the United States is the only Western, industrialized country with significant oil and gas reserves and production.

- Until 1973, the international oil business was dominated by U.S.-based oil companies. Although this is less true today, the U.S. companies remain the largest factor in the industry.
- U.S. companies continue to dominate the refining, distribution, and marketing industries in most countries.
- U.S. dependence on automobile transportation and therefore on gasoline is significantly greater than that of other countries.

These facts mean that both decisions and strategies in and for the United States are necessarily complex. Japan or Spain, for example, with essentially no oil or gas production and with essentially no interest in any anywhere else in the world, are only consumers of petroleum. For them, self-interest lies in protecting their consumers. There is no national constituency promoting the interests of producers.

Policies in the United States are much more complex. The United States is both an important producer and consumer of energy. In addition, an important segment of the U.S. economy has vested interests in the petroleum business in most countries of the world. Complicating matters is the fact that the interests of the United States are highly regionalized. For example, in energy, New England differs from the Gulf Coast as dramatically as Japan differs from the United Kingdom. The Gulf Coast has an energy situation similar in many ways to that of the United Kingdom, and Alaska's energy situation is not unlike Mexico's. One might liken the Midwest's situation to that of, say, Germany.

The U.S. regional differences have led to a few raw nerves. In the heyday of the OPEC cartel, oil-poor regions of the nation watched their manufacturing industries suffer while oil-rich states reaped the bounty of skyrocketing oil prices. With the collapse of prices in the 1980s, the tables were turned; many in the depressed oil patch thought the economic boom of the rest of the nation came at their expense. How policymakers navigate these dangerous shoals is best left to the wily politicians.

Despite the complexities, the "lessons" we offer, which have been learned the hard way, should help to inform policy and decision makers at many levels.

Lesson 1. Depletion is not the issue. The issue is not energy substitution, but technology substitution.

Lesson 2. The consequences of poor strategies can be enormous and unpredictable.

Lesson 3. Most forecasts are wrong; therefore, robustness is a very important requirement in planning.

Lesson 4. Analysts should not attempt to tell the decision makers what to do. The task of the analyst is to lay out the options.

Lesson 5. Measurements may be wrong.

Lesson 6. Do not confuse a systems approach with a systems analysis.

Lesson 7. Understand and manage technologies accordingly.

Lesson 8. Quality pays, and cleaner is cheaper.

Lesson 9. Do not overemphasize science and underemphasize engineering.

Lesson 10. Government-sponsored R&D projects in areas where the government is not the user of the results are usually ineffective.

The lessons should help each region apply the principles of Integrated Energy Systems to best fit its own situation. A national energy policy—if there can be such a thing in the United States—must be one that allows and supports a mix of approaches by each region, for there can be no single solution, but only a common set of lessons, criteria, and principles by which better decisions can be made. This also applies to each region and each country of the world.

## ENERGY AND UNCERTAINTY

Another MIT colleague, the late Prof. Fred C. Schweppe, popularized the wisdom that the phrase "planning under uncertainty" is an *antithesis*. That is, in the sense we use the word *planning*, no planning is needed if there is no uncertainty. If you know what the future holds, then you can schedule your actions without the need to think of contingencies. It is uncertainty that makes planning—with its emphasis on contingencies and risk—so important. Superimposed on the political, economic, and societal differences among regions and countries, uncertainty is unquestionably what makes planning so difficult but at the same time so necessary.

In this book, we have elaborated on the uncertain features of the oil industry: in supply, in demand, and in prices. We further point out the importance of contingency planning. We drew an example for business strategies from the case of gas turbines, and for R&D planning from the case of EPRI. In both cases, we pointed out that the policymakers needed first to identify every contingency they needed to be protected from, although they did not always do so.

In energy, there was no significant contingency plan in the public or private sector before the embargo. We had a plan, but we did not do planning! The only real contingency plan—the strategic petroleum reserve—came too late, was too small, and was too poorly and stingily executed.

Looking at the past, it seems hardly necessary to point out our inability to forecast surprises. In fact the expression *forecast surprises* is an antithesis. Most of the significant energy changes of the last twenty years have come as shocks or discontinuities, each happening within a very short period of time—weeks or months, not years. Each event was discrete, and not part of a trend or pattern. On the whole, we did not foresee the changes in 1973, the formation of OPEC. We did not foresee the social response to nuclear power, air pollution, ozone depletion, and hazardous wastes from industrial processes. We did not have a proper concern for the greenhouse effect. We, in general, reacted to each shock and discontinuity as it occurred. Then, we compounded the problem by projecting from each new discontinuity as if it were the beginning of a new trend.

The central theme of the Global Change Program of the International Council of Scientific Union is that damages to the ecological system from human economic activities may be irreversible and unexpected. "Surprise" is now a most important subject in ecological research. We can't afford any more of them.

Uncertainty is the way of life for planners, and searching for defenses against surprises should be their major challenge. It must become so especially for all involved in the energy sector. The IES approach is one defense against uncertainty. Are there others? Perhaps. The important thing to do from now on is to think about possible contingencies and prepare for them. If we do a good job, the prospects for energy in the future can be bright!

## A RELEVANT POSTSCRIPT

Just as the final work was being done on this manuscript, two of us participated in a national energy crisis simulation workshop. The workshop drew about two dozen energy experts of varying backgrounds to simulate the development of energy policy recommendations to the U.S. president, following "the most serious oil shock in history" in 1993. This kind of research and training is important, and we did, in fact, learn a great deal. Of course, much of the emphasis of such work is on process, since the scenario for 1988 to 1993 with which such a group works must obviously be hypothetical. Nevertheless, the content of the workshop was revealing—and sobering.

The workshop requires each participant to play an assigned role, and work toward a set of specific recommendations. Déjà vu was rampant. It was 1973–1983 all over again. Big energy wanted to put environmental issues on hold. Big oil wanted public lands opened to exploration, higher prices for gas, and tax breaks and subsidies for exploration and production. It was Project Independence reincarnated. The military wanted, of course, first priority. Labor wanted jobs. Many participants saw the solution in renewables and alternative energy sources. Some even prayed for salvation from fusion. Nuclear, of course, must go forward, despite the risks, and so forth.

It was as if we had not learned a thing from our experience.

But the workshop was only a simulation. Perhaps we will have until 1993 to make sure we have learned our lessons. Perhaps we won't have that long. In any case, the one-day simulation convinced us that the lessons to be learned must be taken seriously. It would be unforgivable for us to repeat the blunders; unforgivable, and needless.

A second set of events is occurring as this book goes to press. The academic, business, and governmental worlds are beginning to focus on the issues of global change, specifically global warming and the effect that increased combustion of fossil fuels plays in the production of the greenhouse gasses. The debate has opened, the research potential is clearly there, the opportunity to look forward and backward to identify *new* alternatives is also there. It is discouraging to the authors to note that many of the technologies trotted out for inspection are those we evaluated in the 1970s. Most of the modeling tools that we tried—and largely rejected—in the 1970s are being dusted off, updated to produce environmental residual measures, and running again. There were lessons, as we have pointed out, that do apply to the questions of global warming. Learning implies that the mistakes of the past need not be repeated.

# Appendix
# System Sensitivity Studies

Chapter 7 introduced the concept of an Integrated Energy System and showed, using simple tools and a straightforward analytic technique, the potential for energy systems in which environmental residuals are treated as by-products rather than waste streams. The purpose of the appendix is to carry the analysis one step further. It presents a more rigorous analytic methodology for choosing the structure of an Integrated Energy System, mathematical programming, and a variety of contexts in which an IES might fit.

In an ideal situation, an energy system for an industrial complex would integrate the fuel, heat, electricity, steam, and industrial gas availabilities and requirements for the complex as a whole, *in light of the process peculiarities of each plant in the complex.* The system would operate as it does today, with more integrated economics and additional flexibility.

Chapter 7 discussed the most tightly integrated system package evaluated to date, which includes a set of technologies in various stages of development, ranging from the well-known and well-understood gas separation plant to the less well-understood HTGR technology to the relatively new oxygen-fed combined cycle system (see Figure 7-2). Two significant points emerge from an analysis of the system: its robustness and identification of processes in which small adjustments make significant differences in the overall system and cost efficiency.

The system was evaluated by simulating its performance, given a set of assumptions concerning the external environment. The assumptions were then challenged individually and in groups. Worst-case scenarios were evaluated, knowing well that we might or might not have effectively defined *worst*. The numeraire was mils/kWh only, recognizing as we did then and now (Chapter 5) the fallacy of a single criterion. The self-criticism notwithstanding, the cost of the IES compared to other base-loaded electric power generation technologies was essentially flat.

We also tested the sensitivity of the system to the assumptions made in the analysis. (See Figure 7-9.) Improvements in the performance of the known technology (i.e., the gas separation) made little difference. They were neither the dominant capital structure nor a dominant operating cost. Improvements in the GTCC were more significant, but only slightly. The largest change occurred in the HTGR, which is the technology with the highest capital cost and the greatest future uncertainty.

There are several points of sensitivity beyond the simple mils/kWh indicator that relate to the structure of the system itself. The system "sells" carbon dioxide, presumably to EOR. If there is no EOR, is there another market for the carbon dioxide? Possibly, but more likely the system would be restructured (at least temporarily), not to separate out the carbon dioxide, but rather to vent it to the atmosphere. In addition, it may be more economic to burn the syngas in an air environment than in oxygen. If this is the case, the IES would have fewer components. By the same token it is possible new technologies would substitute in the more complex IES structures. These might include the oxygen-fired gas turbine we have proposed, a new process for oxygen separation such as MOLTOX (Anderson and Dunbobbin, 1985), which is now under development and commercialization, or a replacement for the steam reformation process, the HTGR, and so forth.

Finally, if the goal is to provide services, other combinations of technologies, including those that consume as well as generate or process energy, may enter the system to reduce the overall cost, increase overall efficiency, or meet other societal goals. But how is it possible to screen or to select individual technologies within an IES to best match the needs of an individual country or an individual industrial structure?

Because of its generic configuration, the IES lends itself to mathematical optimization by standard techniques such as mathematical programming. Our studies have revealed that linear and nonlinear programming techniques yield useful and, incidentally, rather elegant results. In the simplest application of such techniques, the demand profile and prices are taken as fixed. The mathematical programming problem can then be formulated to maximize net revenue within the system subject to the set of constraints that would apply. Constraints arise from within the system (e.g., physical laws that govern process flows) and outside the system (e.g., local demand profile). A few examples help illustrate our point.

The discussion below builds upon the information in Chapter 7 concerning the Integrated Energy System. The analysis uses a standard linear programming approach to evaluate the mix of technologies that would be incorporated into an IES. The LP model maximizes the return or total value of all outputs given the costs of all of the inputs, including capital. The inputs to the model are operating costs—which are linear with respect to the quantity of production at any point in the system—and capital costs, which are nonlinear. As an example, capital costs per unit of output increase if a particular unit is underutilized and they increase when a unit is increased beyond what the industry knows to be the optimal size of the production unit. This increase reflects the economics of adding another production unit that will then be underutilized up to a specific point. The description below looks at the relative *structure* of three IES systems given variation in the costs and quantities of input and output. Specifically Scenario 1 is a base case, scenario 2 varies by a factor of 4 the shadow cost of environmental (air) pollution and scenario 3 varies by a factor of 4 the cost of methane.

In scenario 1, the cost of pollution was placed at 5 mils/kWh. In scenario 2, the cost was increased to a very high 20 mils/kWh. The system behaves in an interesting fashion in selecting the optimal configuration in the two cases. The generic IES has incorporated within itself both a conventional air-blown GTCC and an oxygen-blown GTCC. The latter produces no environmental pollutants; it is, however, more expensive to construct because of the need to provide for the recycling of some of its exhaust gas to provide the required gas volume.

In scenario 1 (see Figure A-1) the optimal configuration includes the maximum utilization of the air-blown unit and a lesser utilization of the oxygen-blown unit. In scenario 2 (Figure A-2), the air-blown unit is not utilized at all. This is as we might expect. What is important, however, is the system's sensitivity to environmental cost. In a conventional energy system where the energy flows are separated from each other, environmental effects are externalized. If the shadow cost of polluting the environment increases, the cost of the final product increases proportionately. In the IES concept, it is possible to internalize the cost of polluting the environment. The result of an increase in the shadow cost of environmental effects is the selection of an alternative path that incurs a less than proportionate increase in the cost of the final product. Throughout this book, we have argued that it is better to go with systems that prevent the formation of environmental pollutants rather than cleaning after polluting. As concern over the environment increases and the effects of past abuse become more and more apparent, the conditions of scenario 2 become a closer representation of reality and oxygen-based technologies make more economic sense.

Scenario 3 illustrates the effects of some of the exogenous factors that influence the optimal configuration. In scenario 3 (see Table A-3 and Figure A-3), the cost of natural gas is increased by a factor of four. The result is the redistribution of electricity production to another generating capacity in the system.

**TABLE A-1 Scenario 1**

Scenario 1 was essentially a base case scenario. The resource cost/price assumptions and summary are recorded in the table below. No constraints were applied either on output conditions or on plant size; the model was allowed the freedom to choose the optimum path. Figure A-1 indicates the optimal path allocation. The model choice between the oxygen-blown GTCC and the air-blown GTCC is particularly interesting. The difference between these two systems arises from the environmental perspective. The latter entails an additional cost of cleaning up the $NO_x$ pollutants as a result of using air as the oxidant (shown here as a shadow cost rather than an expenditure incurred). The former involves an additional capital expenditure to overcome some of the technical complications of recycling carbon dioxide to dilute the turbine gas volume. Depending upon the setting of the environmental cost, the model selects either the oxygen-blown unit or the air-blown unit as its first preference.

*Costs/Values*

| | | | | | |
|---|---|---|---|---|---|
| Methanol | $/ton | 250 | Ammonia | $/ton | 150 |
| Methane | $/mmscf | 1,500 | Urea | $/ton | 200 |
| Oxygen | $/mmscf | 500 | Electricity | $/mWh | 30 |
| Nitrogen | $/mmscf | 500 | Carbon dioxide | $/mmscf | 0 |
| Hydrogen | $/mmscf | 1,000 | Coal/Biomass | $/ton | 10 |
| | | | Environmental Cost | $/mWh | 5 |

Return: 27 $mn (total benefits minus total costs)

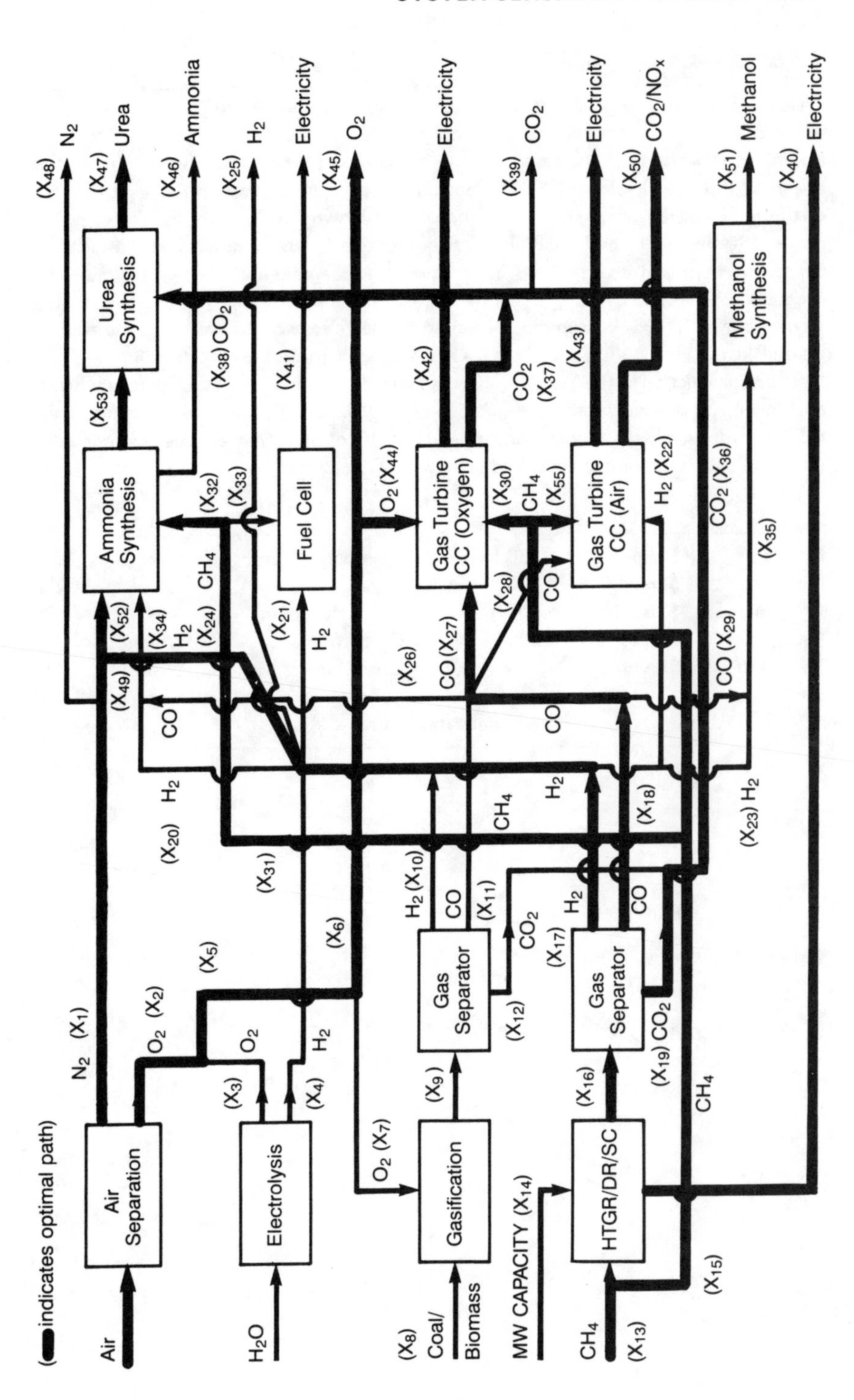


**FIGURE A-1** **Scenario 1: Optimal Path.**

## TABLE A-2 Scenario 2

Scenario 2 was used to illustrate the important distinction between the two GTCCs. In scenario 2, the values from scenario 1 were retained with the exception of the environmental cost, which was increased from 5 mils/kWh ($5/mWh) to 20 mils/kWh. It is obverved that the air-blown GTCC is not utilized at all in this case, reflecting the effect of the increased environmental cost. The oxygen-blown GTCC provides the electrical power together with the HTGR/SR/SC as in scenario 1. Note the manner in which this affects the mass and energy flows in the model. In scenario 1 there was no methanol production. The entire CO production from the steam reformer was supplied to the oxygen-blown GTCC. In scenario 2, because of the release of the natural gas previously utilized in the air-blown GTCC, the CO produced is made available for alternate use. Methanol production has become the alternate feasible operation even though the methanol component parameter values have not been changed. Clearly, the shadow prices of the CO and hydrogen in the model have changed because of the increase in the environmental cost. The path allocation is contained in Figure A-2.

| ***Costs/Values*** | | | | | |
|---|---|---|---|---|---|
| Methanol | $/ton | 250 | Ammonia | $/ton | 150 |
| Methane | $/mmscf | 1,500 | Urea | $/ton | 200 |
| Oxygen | $/mmscf | 500 | Electricity | $/mWh | 30 |
| Nitrogen | $/mmscf | 500 | Carbon dioxide | $/mmscf | 0 |
| Hydrogen | $/mmscf | 1,000 | Coal/Biomass | $/ton | 10 |
| | | | Environmental Cost | $/mWh | 20 |

Return: 8.6 $mn

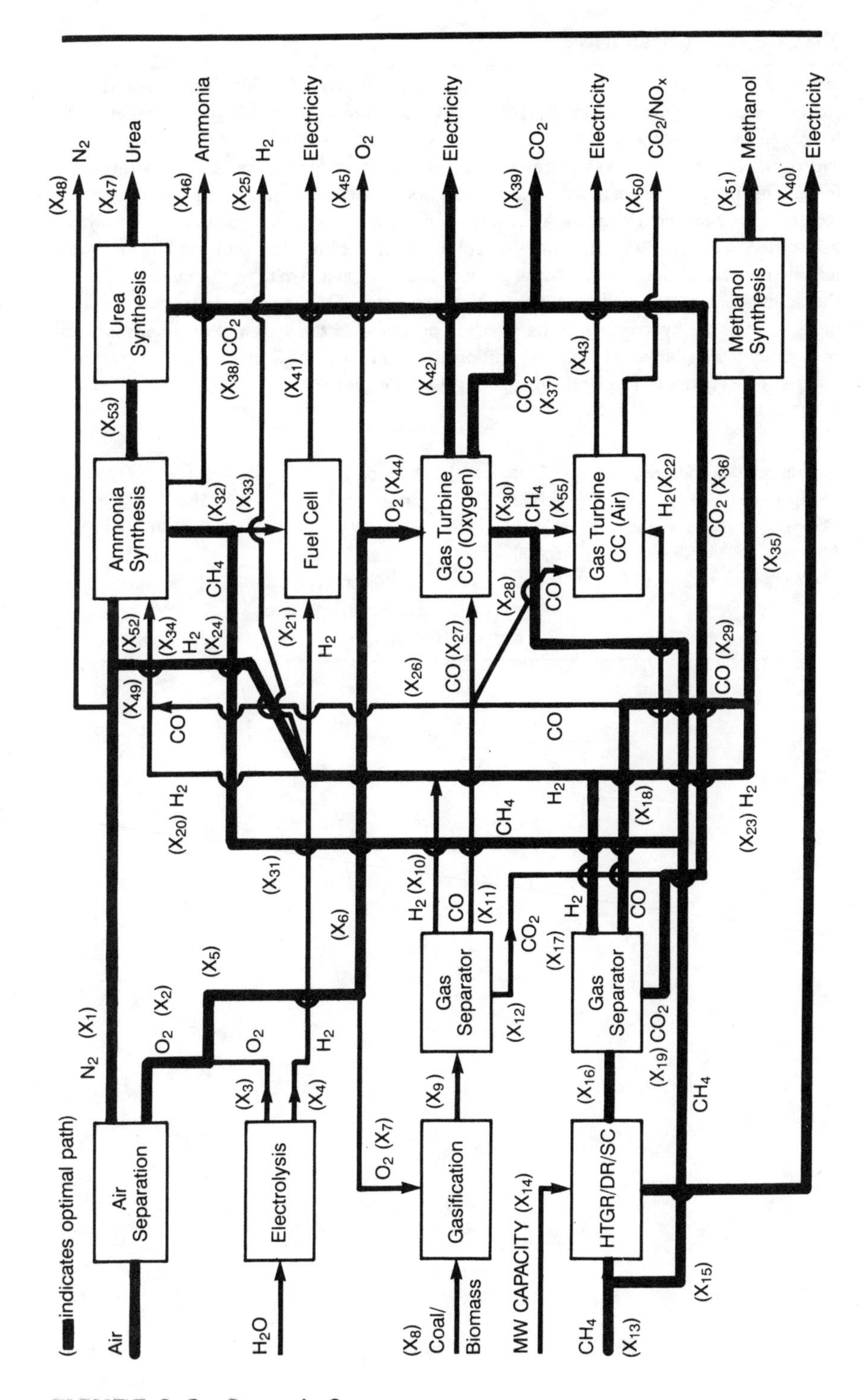


**FIGURE A-2** Scenario 2.

## TABLE A-3 Scenario 3

In scenario 3, the environmental cost reverts to its original 5 mils/kWh. The cost of natural gas is increased to $6,000/mmscf ($6/mmBTU) and the price of oxygen, nitrogen, and hydrogen are also increased, each by a factor of 10. The prices of oxygen, nitrogen, and hydrogen reflect the market prices in the United States for the gases in their end-use form. The model thus reallocates its production in order to maximize the output of oxygen, nitrogen, and hydrogen. All sources of these gases, viz., air separation, electrolysis, gasification, and HTGR/SR/SC, come on stream. Methanol production is no longer found feasible because of the higher shadow price of the hydrogen. Similarly, the model bypasses the ammonia synthesis and urea production processes. This reflects the higher shadow price of both the hydrogen and the nitrogen produced in the system. Power production reverts to the air-blown GTCC (in addition to the HTGR/SR/SC production). This is the same as in scenario 1. The optimal path is given in Figure A-3.

***Costs/Values***

| | | | | | |
|---|---|---|---|---|---|
| Methanol | $/ton | 250 | Ammonia | $/ton | 150 |
| Methane | $/mmscf | 6,000 | Urea | $/ton | 200 |
| Oxygen | $/mmscf | 5,000 | Electricity | $/mWh | 30 |
| Nitrogen | $/mmscf | 5,000 | Carbon dioxide | $/mmscf | 0 |
| Hydrogen | $/mmscf | 10,000 | Coal/Biomass | $/ton | 10 |
| | | | Environmental Cost | $/mWh | 5 |

Return: 100 $mn

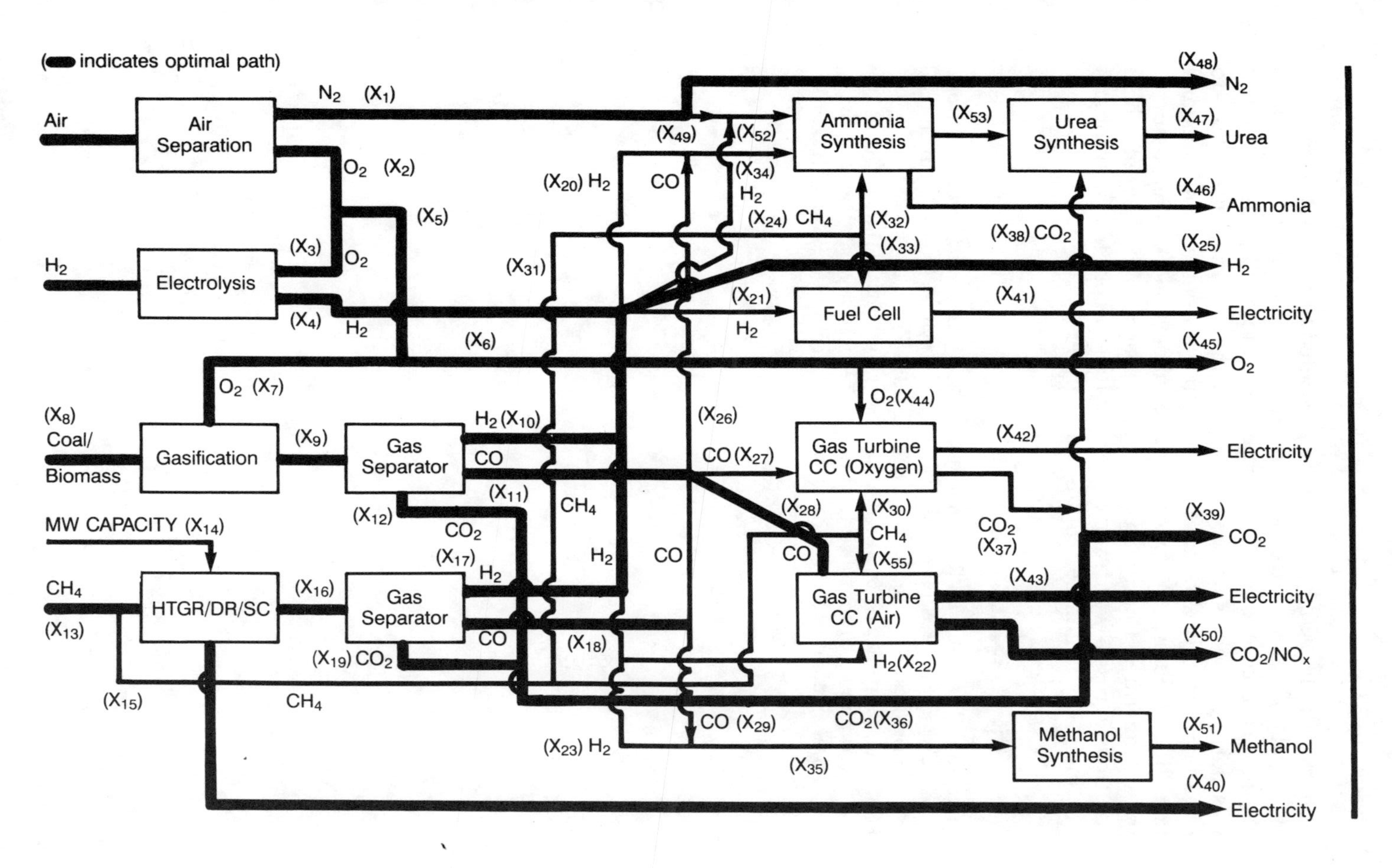


**FIGURE A-3** Scenario 3.

In Chapter 5 we discussed the pitfalls of models and assumptions not challenged. Here we have presented the results of a number of models of Integrated Energy Systems. The reader may want to challenge our consistency. It is important to remember that our point was not that models are not useful, but that we must learn how to use them properly. We have challenged both our assumptions and modeling structures. With our simple models and explicit assumptions, Integrated Energy Systems do appear to offer a set of robust solutions across a wide range of assumptions about the environment in which they would be required to operate.

## REFERENCES

Anderson, R. A., and B. R. Dunbobbin. 1985. *Pilot Plant Development of a Chemical Air Separation Process.* Allentown, PA: Air Products and Chemical, Inc. Report No. DOE/CE/40544-T1.

Fernando, Chitrupa. 1986. "Integrated Energy Systems: A Mathematical Programming Framework for Energy Policy Analysis." Master's thesis, Department of Management and Technology and Policy, MIT.

# Index